HIGH-TECH MARKETING SIMPLIFIED

AN INSIDER'S SHARED EXPERIENCE

TED MARENA

Pro Indie Publishing

www.proindiepublishing.com

Book cover designed by Terri Dilley

Author image photographed by Two Dudes Photo

Book formatted by Jennifer Eaton

As I was finishing this book my father was fighting for his life. To everyone who is battling illness or diseases I am dedicating this book to your struggle, perseverance and will to live. Positive thoughts and wishes for a speedy recovery and blessing to you all. Encourage those who are not feeling well and support them as best you can. Heed this lesson as a reminder to take advantage of the moments in front of you. Imagine if you were in a coffin at your wake and friends and family were paying their last respects. How many of them will say I'm so glad you completed all the work tasks you were supposed to do?

What you want them to recall about you are memories, stories, meals, experiences, moments, and so on that you enjoyed together. Make the time to enjoy life and spend time with your family and the ones you love. I'm glad my father and I shared many moments like our trips to Italy, Canada and several places in the USA. We had many holidays, events and meals shared together even when I lived on the other side of the country. These were the moments I focused on at his wake. Let's not take anything for granted. Your life will be richer and more fulfilled.

CONTENTS

PREFACE

The reality is there are numerous types of marketing: digital, strategic, product, promotional, tactical, etc. This book is focused on the critical steps required to define, position, message and launch a technical product, such as a semiconductor device. All of the key experiences, across numerous companies, which I've learned are shared in *High-Tech Marketing Simplified* to help you understand what marketing is required from start to finish. Learn the insider secrets to improve your odds of success.

Everything I have experienced in my extensive career in semiconductors and various technical roles is in this book. *High-Tech Marketing Simplified* is entirely based on my technical background, extensive marketing and sales experience. This book provides specific details for technology products and expands upon the 7 steps I defined in my previous book, *Marketing Simplified: An experience based step-by-step guide to grow sales.* I've written this book to share the additional details required for technical and semiconductor marketing and business development. More than just a guide on how to execute a technical product promotional campaign, *High-Tech Marketing Simplified* encompasses marketing activities from product inception to customer sales and support.

This book is not leveraging any research or university study; I'm drawing on more than 25 years of electronics design, technical support, sales, business development and marketing experience. If you are looking for all the details needed to effectively market your semiconductor, technical product or service, this book is for you. Given that most of my career has been in semiconductors, much of the focus will be for that technology, but the majority of the content is applicable to other technical and related fields. There is no pontification or needless background data to explain the key actions and process steps required. This book gets right to the point, and I use many work examples to further clarify each stage of the process. *High-Tech Marketing Simplified* is intended to explain, in an organized step-by-step manner, how to define, segment, position, go to market and promote your technical product or service. There are ten fundamental tasks to follow through the process, and I've laid them all out for you.

If you are a semiconductor or technical professional supporting product or technical marketing or marketing communications and you want to improve your skills, this book will prove invaluable in that endeavor. The process laid out will be a priceless guide to follow for future product releases. In addition, start up companies, technical agency representatives and consultants will benefit greatly from the information presented. C level executives and marketing leaders will also gain new insight into what is really needed in order to achieve successful technical product launch and execution. There is something for everyone in *High-Tech Marketing Simplified: An Insiders Shared Experience.*

INTRODUCTION TO HIGH-TECH MARKETING SIMPLIFIED

YOU DON'T JUST DO MARKETING

Before we begin, I want you to recall a sales or marketing campaign that you think is memorable, creative or impactful. Is it a Super Bowl commercial or an Apple product reveal or some social media advertisement? Whatever you are remembering and sensing is likely the same impression and feeling that you will want to create when you market to your customers. By implementing the *High-Tech Marketing Simplified* process steps I will lay out, you can achieve that result. Will you change the world? Most likely not, but your technical product, device or solution will be improved by applying the insider knowledge I'm sharing.

I find it peculiar that so many people think they can do marketing. I've heard from numerous individuals who seem to think marketing is easy to do and that no training is needed. Perhaps it's because you don't have to have excellent mathematics skills or know programming languages or possess some other technical background. The reality is marketing is a proficiency that has to be mastered just like any other career skill.

The good news is that I've simplified what you need to do to define the markets to target, create a go to market plan and determine how to best promote. Let me share a real story first just to explain why this book will be valuable to you.

Figure 1. Marketing is not easy, but with this book, it is simplified, and you can do it!

A couple of years ago a technical nonprofit organization, RISC-V (Reduced Instruction Set Computing – 5th generation) International needed a marketing leader. RISC-V is an open instruction set architecture that is used to define a Central Processing Unit, CPU. The group's marketing chair was going to be elected by the member companies of RISC-V. The individual who won the election was a software engineer. He ran for the position because he thought it would be fun and easy to do. I was surprised to find out what little direction and knowledge the new marketing leader had about what marketing professionals should do.

I supposed I should not have been surprised given he was a software engineer. After all, I didn't know how to program in C or Python software languages either. What was clear was that he had run for the position not understanding what was required to be successful. After we worked together for a couple of months, he stepped down and then suggested to the RISC-V organization that they appoint me the marketing leader. This story was not shared to proclaim only marketing backgrounds are qualified to do marketing. The point is that marketing is not something you approach as an afterthought. Don't simply go through typical marketing activities and call the job done. I was inspired to write this book because of the feedback I received after my first book *Marketing Simplified, An experience based, step-by-step guide to grow sales.* That book was targeted more generally and aimed to help those with less marketing experience. Given that I've been in numerous technical and semiconductor roles, I wanted to share that specific and detailed knowledge with you. If you want to know what insiders learn over numerous years, or you just want to improve your marketing and sales capabilities for technical products, keep reading.

This book will help you become a better marketer and improve your technical product or service. More specifically I will share the fundamental marketing tasks and the detailed process required for semiconductor and high-tech products or services. Whether you are an experienced technical marketer or just getting started, the information I'll share will be a great guide for you. It doesn't guarantee success, nobody can do that, but you will absolutely improve your odds of success if you follow the High-Tech Marketing Simplified process steps.

LEVERAGE MY EXPERIENCE
SO YOU'LL MAKE FEWER MISTAKES

Often high-tech companies want to differentiate via their technology or innovation. While this is desirable and often needed, few think of using other avenues to distinguish themselves. I submit that companies should also be differentiating with marketing! In the current environment, it is easy for companies to copy products, software, apps, etc. There exist more tools than ever that can be leveraged to develop similar offerings. There is no better example than Instagram. With only 13 employees and a handful of investors, they were able to use software tools in the cloud, leverage web based resources and grow their user base to hundreds of millions.[1] Even though Twitter, Facebook, Myspace and other social channels were firmly established at the time, Instagram was able to duplicate and improve on the existing offerings with very limited resources. The moral of the story is that when products offer similar functionality, marketing is what can help separate the winners from the losers.

What will be shared with you is all based on my actual career experience working in many high-tech marketing roles, business development, technical support and sales positions. I've seen first-hand that products with supposedly the best features and capabilities do not always win. If you are an engineer, you might be thinking that I'm crazy. The reality is that the best product is not always the winner. Let that sink in for a minute.

Of course, a great product or solution is important, but marketing is an essential factor in how the product is perceived and experienced. There are many marketing facets I'll be walking you through. I've laid out the process necessary to execute a high-tech marketing and sales campaign from start to finish. These marketing activities will increase your likelihood of success for your semiconductor device, technical product or service. One last diversion before we jump into the High-Tech Marketing Simplified process, I want to share with you one of the reasons I excel at marketing.

If you know me well, then you are aware of what the number 410 represents. No, it is not a Chinese number sequence or anything culturally related. 410 represents the score I earned when I took my English SAT test. The SAT was a standardized test taken just before university enrollment. 410 for an English score is not considered a good score by any measure. One of my good high school friends regularly reminds me that they give you 200 points just for signing your name. I always struggled trying to understand or construct complex English phrases. Why am I sharing this supposed weakness of mine with you? Well, this "weakness" is one of the reasons why my marketing programs have been successful. Whenever I communicate, I always keep it simple, because I have to! Big, fancy words or phrases were not something I could relate to or easily understand. I am always looking to simplify sentences, simplify messages, simplify technical instructions and so on. If a message is simple, people will understand it and become interested to learn more. I hope my clear communication in this book will allow you to absorb the material easier.

Of course, simplifying your message and story is important, but hardly enough. Let's jump into the High-Tech Marketing Simplified process and explain each stage briefly. There are ten key stages in the process. Some will be easier and more natural for you than others, but all can be accomplished. First, we start with selecting your segment. We'll discuss why this is important and how it can be the difference between success and failure. The next stage is positioning your product. For this process step,

I will define why a complete solution is critical before you begin promoting your device or product. Next is goal setting. Believe it or not when you are marketing, it is absolutely important to have goals. It is essential to set at least one goal. Then I'll explain messaging in detail and how to set up an impactful top line message for your technical product or service.

Once you have the messaging set, you will define your go to market details: how and where you will be selling your solution. Then you will create a map or plan detailing what you will actually do to support the go-to-market. The next process step is creating your collection of marketing material, also known as collateral. What specific collateral you need to create will be defined when you map out what you will deliver and where. All your collateral will echo and amplify your messaging as we will see later. The next step is what I often referred to as "herding the cats". Consider this aligning and communicating with the organization to ensure everybody is working toward the same goals. This is important if you are in a sizable organization or larger company. If you are in a smaller company, then this step will be quicker for you. Then comes the delivery and execution stage: what you're actually going to do, how you're going to do it and who is going to do it. The last step is measuring your marketing efforts. All marketing campaigns should be measured against the original goals that were set. By doing this you will be able to make adjustments and continue your sales progress.

To summarize, here are the stages of the process to follow.

1. Selecting your Segment
2. Define your Positioning
3. Setting your goal(s)
4. Messaging
5. Develop the Go To Market

6. Mapping out your delivery
7. Building your collateral
8. Herding the cats
9. Delivering
10. Measuring

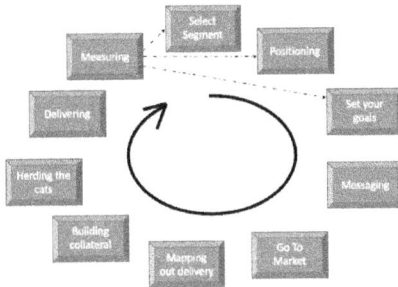

Figure 2. Visual representation of the High-Tech
Marketing Simplified process

Before continuing, I want to emphasize that the process of what you do is more important than all of your other efforts. The steps I have laid out here are the process I am proposing you follow for improving your marketing capabilities and results. I've often heard that setting goals is the most important action you can do, but I disagree. It is the process that is the most important aspect to follow. The most successful college football coach in America is Nick Saban. He regularly talks about goals being important, but not as important as the process. Nick Saban has repeatedly said, "Focus on the process of what it takes to be successful."[2] His experience has resulted in numerous winning seasons and many championships. The process is what you can control, execute and implement. The ultimate outcome and results are often out of your control, but what process you follow is within your control.

The high-tech marketing process, when followed, provides you a framework which allows you to make adjustments after you measure your results. I've found these process steps to be important to best launch a product. To recap, creating a product or setting a goal is great but not nearly enough. Most technologies are somewhat better or differentiated but not to the point of being irreplaceable. Competitors can duplicate or copy your product faster than was previously possible. Use marketing to improve upon your position in the market and in the minds of the customers. Let me explain how you do this by executing the process of the High-Tech Marketing Simplified steps.

2

SELECTING YOUR SEGMENT

DIVIDE AND CONQUER

Before we jump into segmentation, let me provide a distinction between a product or device and a solution. A product could be a device, such as a semiconductor chip, a software program, or a box level product. A solution on the other hand are all the items which support the product. The solution contains the pieces that are needed for customers to easily adopt your product. Think of the solution as the larger set of items that make up the ecosystem for your product or device. In the next chapter we will explain solution details further. For now, understand that there is a difference between a device or product and a solution.

TIPS TO DEFINE YOUR PRODUCT FEATURES

It is likely your company has an existing product or is thinking of a follow-on device, or maybe you have an entirely new product idea. When thinking about product features, listen to customer feedback, but you also need to anticipate what their needs may be in the future. Some customers will not know future requirements, but others may. Also, just because a customer does not ask for something, does not mean you should not consider adding it. Obviously adding features is a risk as it drives up cost, especially if nobody every uses it. However, additional features could be a differentiator. Think back to 2007, do you think anyone asked for a full screen mobile phone before the iPhone was introduced? The answer is no. Customers were not aware the technical capability existed, but once they found out about it, they all wanted it. Blend your technical experience and organizational capabilities with customer feedback when defining product feature requirements.

Occasionally product features you never realized were needed will suddenly become key requirements. Sometimes it is because the market moves, or competitors introduce these capabilities, or other factors influence the customer needs. You need to react as efficiently as possible when market shifts occur. It is best to balance the direct feedback from customers with your market knowledge, the technical possibilities, and your organization's capabilities. If you build the same exact product as your competitors, then you will have to sell more on price than differentiation. You may still be able to win with a very similar device if you can leverage your company brand or if the organization has many other products that customers already use. If you are a smaller company, look to differentiate in a couple of ways.

How do you differentiate? We are assuming you have reasonable knowledge of the customers and product arena in which you want to compete. At a technical level, after you have included all the key requirements that the majority of customers expect, then you need to add, subtract or upgrade features. Earlier in my career, we introduced a product that could be programmed and reprogrammed while the device was on the board and in the field. Before this capability, all products had to be programmed externally in a programming machine and then the board was assembled with the programmed device. This new technology upgraded what the customer could do. This additional programming flexibility allowed our company to penetrate a new market because it was differentiated from the competitors.

Another way to look for additional features is to see if you can integrate adjacent products. What is it that your product connects to or talks to? Sometimes you can consolidate other devices and provide higher levels of integration. As an example, there was a mobile product we were selling to process sensor data. Our solution provided the processing needed at a low power level. We unfortunately lost the business when a vendor who sold a sensor integrated the processing we were doing in their new sensor chip. For the customer, this meant fewer devices and lower total power consumption. These are the reasons why our solution lost. Understand what you connect to or might be often used with your device. If you have access to that technology, you should consider adding it. These may be additional capabilities that make sense to incorporate.

If the product you are introducing is truly considered a commodity, that is fully replaceable, you should still strive to differentiate. Can you produce the device with lower power consumption? Can you make it faster? Or think about producing something that is easier to switch to, has lower latency and so on. The point is to differentiate in a way that adds value to your customer. The easiest thing to do is offer a lower price, but that should be your last action.

Once you have your product defined and created, you need to select a segment or a few segments to focus on and win. Ex-Intel Corp. executive, William Davidow, once said, "Marketing is civilized warfare."[3] Harsh language? Perhaps it is, but you need to determine in what parts of the market you can be successful and, more importantly, those in which you will not.

For technology companies, you can think of a market segment as a collection of customers who have a similar need or requirement. What makes it somewhat difficult to identify a segment is that the customers are not holding up signs saying what sector they are in. As a marketer and company, you must determine which types of customers care most about what features and capabilities. For some market segments there will be similar needs and requirements, but there will also be differences. It is necessary to determine the dominant needs of the customer population. Some companies will have a few different requirements than most other segments, yet they may also have needs that span multiple segments. In the end, you have to analyze and boil down what are the key factors that you see customers demanding that most closely match your product and company capabilities.

GO FOR THE BIGGER FISH IN A SMALLER POND

You need to adequately fund what is necessary to be a major player in which ever segments you chose. Often it makes sense to limit the number of segments to 2 or 3. It is better to focus on dominating a smaller segment than trying for a smaller piece of a much larger market. This is especially true if you are a newer competitor and if you have fewer resources than the larger companies in the bigger segments. Whatever segments you select, you need a marketing strategy that will enable you to win. By winning, we're talking about a segment that you can be number 1st or 2nd, or very worst 3rd in. Don't try to be a small fish in a big pond.

Start by dividing up the entire market and look at the segments in which you believe you can excel. Match the needs and requirements to those of your capabilities and resources. For example, in the automobile industry, if you are a small automobile manufacturer, you don't set a strategy to win and fight in every area of the business. You segment the overall business, such as small SUVs, luxury cars, electric vehicles, sports cars, trucks, and so on. Pick one or a few segments that you can support completely. It is better to capture a larger share of a smaller segment than a smaller segment of a large market. The smaller suppliers in bigger segments are often the least profitable because they cannot scale and can't fund the necessary infrastructure and ecosystem needed. In many larger markets, the majority of the revenue goes to a small number of larger suppliers.[4] The smaller suppliers are left to fend for the scraps.

Many companies often underestimate the cost it takes to enter larger markets. In the early part of my career, I was in a company that wanted to compete in the entire Field Programmable Gate Array, FPGA market. At the time we were only in the lower end of the market and doing quite well. The company was profitable and had a loyal customer base. We made huge investments to attack the high end of the market. We developed products that were differentiated and compelling. What we never realized is that having a slightly better product was not nearly enough to win. Our software tools were nowhere near the capabilities of the larger, well entrenched competitors. The solutions which we created to showcase our products were inadequate. We'll talk in detail about solutions in the next chapter. The bottom line was that it became very clear in a short amount of time that we were not going to win in the high end. You can't just introduce one product and think everyone will adopt what you introduce, no matter how much better it is. Not only do you have to make the first product, but you also have to follow that one up and start developing the next one. In addition, all of the ecosystem around the product also has to be developed, documented, etc. In the end, we did not realize these specific details because we were not actually in the high end market.

Smaller organizations and companies that are newcomers should focus on smaller size segments. Sometimes the various segments are referred to as vertical markets. Whereas the broad market characteristics are considered a horizontal market. For example, a vertical market could be automotive, medical, consumer, industrial, financial or government. As mentioned above, a horizontal market would be the entire FPGA or the CPU market. Horizontal segments could also consist of features, such as security. Segments such as government, medical and many more are always concerned about high security. Since security is a feature many different vertical segments are interested in, its wide appeal makes it a horizontal need. The key capabilities of your product should help you determine the vertical or horizontal markets where the feature is most desirable. Lastly, recognize that no segment is completely universal. For example, many consumer customers may want low power devices, but not all of them absolutely need it.

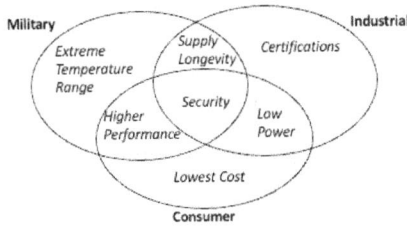

Figure 3. Horizontal versus vertical segments

When you focus on a particular or a limited number of vertical markets, the organization can better target its resources. You can concentrate on delivering the necessary and key customer requirements. It will also enable the company to beat larger competitors who likely cannot provide the same level of focus. Often larger companies are concentrating their efforts in the larger vertical segments and the broad horizontal market. It is not that a smaller vertical segment is being completely ignored, but these larger companies are counting on their brand, breadth of product, having a broader ecosystem, etc. to address the smaller vertical segments. Smaller organizations or new entrants will have an increased chance of success when they choose a limited number of vertical markets.

If you have a good product and are making progress in your chosen segments, then you can look to expand the vertical markets to compete in. Let me share a personal example. I worked in a company which wanted to grow a recently announced product family. The product was doing well in the defense and military markets where it was primarily targeted. During my interview process I talked about how this product should also be selling well in the industrial and IoT space. Obviously, they liked my thinking because I was hired to do this. After I started, I learned that most of the people in the organization came from military, government, and defense markets. This was the main reason the product was originally focused on those vertical segments. Making the product successful in the industrial and IoT segment required significant effort, but ultimately, we made it happen. I'll be speaking more about this in the next chapter when we discuss product positioning.

To summarize, once your product is defined and your segment is chosen, you need to focus the energy of the company to win in those vertical markets. There will be many distractions and other opportunities that will look like home runs. Be careful not to divert too many resources going after business that is outside your key vertical segments. Think of your most important segments as land on a battlefield and you will do everything you possibly can to win that ground and defend it until your last breath.

3

DEFINE YOUR SOLUTION FOR POSITIONING

POSITION YOURSELF OR THE MARKET WILL DO IT FOR YOU

Now that you have decided on a product and segment, you need to determine what is needed for complete customer satisfaction. Let me further clarify what the difference is between a product and a solution. When we talk about a product, we mean a device, such as a semiconductor chip or a box level product. A solution on the other hand includes the device and all the other items that are needed for customers to easily adopt your product. For example, a configurable semiconductor power device requires software to program it, data sheets, design examples, cables, application notes, evaluation boards, how-to videos and more. The collection of the items needed to support the device are referred to as collateral. All of the collateral that is needed and the device or product is what makes up the complete solution. The key is to define and create whatever is required to reduce the barrier for a customer to adopt your product. A comprehen-

17

sive solution is when you can completely satisfy the customer and win their business. Moving forward we may interchangeably use device, product or solution when referring to the various marketing topics but realize there is a difference, as previously discussed.

The solution you define and create will be based on your selected market segments. Some customer requirements will be the same across vertical segments, but others will not. As stated earlier, low power, high performance and security are characteristics that many different segments require. If you determine these features to be the case for your chosen market segments, then you need to create the collateral for each of the characteristics to support the customer's needs. As best you can, you have to determine what collateral or collection of deliverables is required and which are nice to have. If you are currently in the market with an existing product, many of the basic requirements will be known to you. Determine the list and plan on developing all the collateral that you can. At a minimum, all of the required items must be obtained or developed. Sometimes you will not be able to fund a complete solution, but you should if you have the resources and money. If not, prioritize the most important pieces of the solutions and develop those fully.

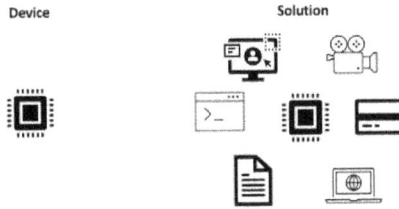

Figure 4. A device versus a complete solution

With your selected segment or segments to focus on, your solution should reduce the barriers for customers to move to your device. When you are creating a solution, do everything you can to simplify the adoption and make it attractive for others to connect to it. Again, having only a device or product is not enough. You need to demonstrate a solution that solves your customers issues. Ensure you create all the necessary collateral whether it be software, drivers, documentation, training guides, evaluation kits, forums, reference designs, etc. The weaker your total solution offering is, the lower the price you will be able to charge. Companies that are leaders in segments did not just wake up there. All the additional collateral that supports and improves on your device is necessary to be a leader in a segment. When your solution is comprehensive, you will be able to command a higher price. Many companies fail to recognize that the costs to create a complete solution are often more than the cost to develop the device or product. These additional efforts are needed if you want to win and support better product margins.

To reiterate, when you pick fewer segments and select them carefully, you can focus your resources. Another benefit of picking a segment is that it tends to simplify the product offering, making your breadth of products and solution deliverables less complex. For example, if you are selling only into customers who want the smallest possible form factors, then you would package your device in the smallest possible size. Larger packaging does not need to be offered. By keeping the number of products focused, allows you an opportunity to better service and support that customer segment. If you were to go after the broad market, then you have multiple packages, numerous software versions, more documentation, additional testing, etc. The additional complexity is often a recipe for disaster and failure. To reiterate, select one or only a limited number of segments to focus on.

As you create the collateral that makes up your solution, always be thinking about how to differentiate your solution from the competitors. It is likely that the device you offer is differentiated from the device you are competing with. Figure out if there are unique methods to highlight the differences. Often these differences are in the details that only product experts understand. Think about how to simplify these differences so it is easy for customers to learn about them. Realize that many customers are overwhelmed and do not have the time to learn all the specific details of each offering. It is likely they already know a competing solution. You need to make it easy for them to learn the key benefits of your offering. As you create all the items that make up your solution, think about how to differentiate it further from the competitor and make it easier for the customer to understand how your differences are better for them.

An important consideration as you create your solution collateral is to show proof points. What I mean by proof points are data-based results on your device. For example, if there is a common algorithm, industry benchmark or likely scenario in which designers will use your device, then perform that function and publish the results. Engineers are influenced like everyone by emotions, but they are even more reliant on hard, concrete data. Ensure that your solution includes results from some common scenario or benchmarks. Documented correctly, these can be very powerful. For example, when I was selling complex programmable logic devices, CPLDs, one competitor who ran the fastest on a new benchmark, published a sales comparison showing their products versus the competitors. Shortly after they did this, every engineer I talked to would suddenly begin to ask how we performed on the benchmarks. It was a difficult objection to overcome even though most customers did not implement exactly what the benchmark performed. The lesson learned was that proof points can be powerful. At a minimum, include results that you can document, and develop a white paper on how you came up with the result to back up your claims.

I can't emphasize enough the importance of delivering as complete a solution as possible. It is frustrating when companies think they can pursue additional segments and more activities than their resources allow them to do effectively. Companies often underestimate the time and money needed to effectively compete in a market segment. If your organization is resource constrained, do a limited number of things better than anyone else for your solution. You must create all the minimum required collateral such as a data sheets, how-to videos, application notes, etc. After the required deliverables are created, develop as many of the other items to ensure a solution is as complete as possible. A great example of a complete solution is the Arduino or Raspberry Pi ecosystem. Both of these hardware boards can be considered the device. They are however, much more than just a low cost board platform. Each has example software packages, support forums, expansion board options, multiple 3rd party contributions and much, much more. Do what I often tell my kids. Put in all the effort required and finish the job the right way. The last 20% always takes 80% of the effort, but it is necessary if you want to win more than your fair share in the market segment.

PRODUCT POSITIONING

I talked extensively about creating the best solution possible because it critical to how your product will be perceived. With your solution defined and your segments chosen, you now need to position your product. In my opinion, the importance of product positioning is directly related to how profitable your product will ultimately be. Position your product and solution or the market will position it for you. The organization needs to determine where they want the product to be perceived relative to the key customer considerations. Product positioning emphasizes the benefits of the solution to your target customers. Further details of what medium or channels you want to use for promotions will be explained in later chapters. For now, think about what the key characteristics of the segment are, and determine how to best promote your product and solution to spotlight them. Consider how and what you want to highlight about your solution that will resonate for the chosen market segment.

However, just because you want to position your solution in a particular way, does not mean you can if it is not backed up. For example, if a memory company wanted to position its product as the biggest size or largest density, then it better back it up. The company has to deliver on this positioning, otherwise customers will think differently. Again, the market will position your product or solution if you don't pick the right attributes on which to promote and focus on. Larger companies may have the budget to pay for research, surveys, or study groups to determine the key influencers that customers demand. Often, simply asking some customers what the most important product attributes are will unveil adequate answers. Understand that positioning goes hand in hand with the chosen market segments. Customers in each segment will have certain characteristics which are most important to them depending on their product needs. Make sure you take this into account when you think about your positioning and how to promote your solution attributes.

Earlier we discussed the various collateral needed to create complete solutions around your product. I want to share with you the product positioning change I led at a previous company. As I mentioned earlier, I went to work for a company with a product selling well in the military and defense market. My initial proposal to additionally position the product for the industrial and IoT market was not welcomed by everyone. Almost every salesperson and business development individuals were telling me the product is only competitive for military and defense. I was told it is dead on arrival for all other markets. In fact, most individuals told me they were pre-selling the next generation product, even though it was two years away or more! Undeterred by the negativity, I plowed ahead believing we could succeed in the industrial and IoT markets.

To reposition a product means you must make changes. First, the organization had to be convinced to update the product messaging. In later chapters we will discuss the importance of messaging. For this product we focused on selling the lower power, abundance of system on a chip, SoC features and product supply longevity. Our new positioning did not push uncompromising security and high temperature ranges which were the key positioning points for military and defense. Instead, we positioning the product as offering more features and lower power, which the IoT and industrial markets wanted. The positioning was to promote the additional features and sell the value which included lower power. Next, we put together a list of deliverables needed for a solution to effectively compete in the industrial and IoT markets. These were items like lower cost and simpler evaluation boards, different application examples, updated product flyers, etc. All of these were important to make it a more complete solution for the new market segments.

As we developed and built the new collateral, we also started to assign goals. I talk in more details about the importance of goal setting in the next chapter. For this new solution, we set a goal of design-ins because that is the start of the journey to ultimately lead to revenue. As we began to promote the new positioning, we refused to allow the sales team to tell us that the larger competitors had better features or other attributes. We focused on what we did well. For example, our longevity record was second to none, and this was important for many industrial customers. The rollout was not an instant hit. We had to engage the field over time with new collateral, hands-on training, joint customer visits and so on. The sales team was reluctant at first, but they saw we were determined to be successful. Ultimately the salespeople saw that they could sell the product in the industrial and IoT segment. From then on, they were bought in and helped us fight for more than our fair share.

One visual tool you can use to think about your product positioning is a 2 x 2 matrix diagram. This is a tool often used to make decisions but is also helpful in seeing how you position your product or service relative to the competitors. Typically, on one axis you will have price. This is displayed in the diagram which has increasing price from left to right on the X axis. The other axis could be features, capabilities, density, etc. You should pick an important characteristic or a few capabilities which are common for your market segments. In the diagram on the Y axis, you will see increasing features as you go from the bottom to the top.

Each quadrant of the diagram roughly defines the type of product that a typical customer would observe. For figure 5, in the lower left quadrant will be lower-cost, less-featured devices. In the upper right box will be full-featured products and solutions at higher costs. You should place your competitors' product in the locations that are most appropriate and position your product as well. Be honest when you locate the competitive products and where you want to position your product.

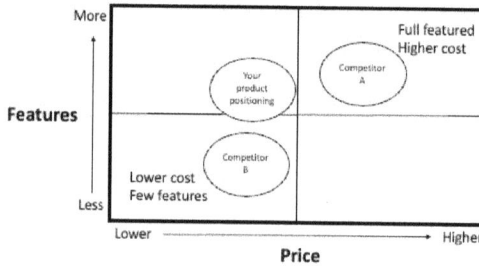

Figure 5. Two x Two Positioning Matrix

Now observe where your product or solution is relative to where you believe the competitors are. Are you directly on top of or very close to a competitor? Sometimes this is unavoidable, but often you can modify the positioning of your product. When you are further away from the competitors, it will be clearer for customers to understand where your product fits in the market. The 2 x 2 positioning matrix is a good tool to visualize and help communicate with others where you want to position your product and solution. You can make multiple 2 x 2 matrix diagrams if you think there are more than a couple specific features which are critical for customers. Again, this is a tool to help you visualize where you will be positioning your product and solution. Recognize this diagram is not an absolute reality, but it should help you clarify your positioning.

It is not always apparent, but product positioning involves

more than the solution. It can also include service and support as key elements. Think about all the resources that you will bring to the segment, whether you have field application engineers, (FAEs), technical forums, instructional videos, samples on demand, mobile application, etc. Other than the device and the solution, what else is a factor for customers in the segment? Also, think about creating a flow chart of the service you offer. For example, what will the customer do or who will be called on if something goes wrong. You may as well let customers know that you are prepared to support them. Usually, all solutions have issues, and if the customer can't fix it, they at least know to whom it should be escalated.

If the support collateral in your solution is complete, then customers will view your service as superior. Often companies with poor documentation require more employees to support the customer. Although customers may prefer to utilize a person for support, this model is not easily scalable or cost effective. For very large customer accounts, be prepared to offer higher levels of support service. These could be supply chain monitoring, forecast sharing, quality reviews and more. Although the costs to the organization increase, the company is likely to become more important for the customer. In time, the customer may become dependent on you. After all, the customer has to invest in people on their side to build the relationship as well. As technology matures or commodity products are sold, service is often a differentiator. Remember you are serving the customer!

CORPORATE POSITIONING

Positioning of a product and solution should be aligned with the corporate positioning. The corporate positioning is likely a reflection of the culture and history of the company. Just as in product positioning, customers position companies based on their experiences, perceptions, what they read, see, etc. The culture and history of a company is how customers often experience what the company is about. The product positioning does not have to be exactly the same as the corporate positioning. However, it should not be the opposite either. For example, the product may be positioned as the lowest total power and easiest to use. The corporate positioning could be a high-tech leader or innovative and efficient technology player. As long as the product and solution positioning are not completely opposite, then you will have a reasonable chance of driving that perception through the customer base. Above all, you must deliver on whatever your corporate position is.

Corporate positioning changes are often tied to new product introductions perhaps because of new technology or market shifts. When companies decide to extend their product offering into new areas, this is often a huge challenge. Not only must the product succeed but the organization must grow into this new positioning. Moving to an adjacent type of product, such as a higher-end device or a complementary product requires the entire organization to invest and learn all the details required to be successful. The effort and cost required to make this product and corporate transition is often underestimated by many individuals in the company. You enter into a new segment with excitement and enthusiasm, but you soon realize the enormity of what is required when you lose deal after deal after deal.

Early in my career, the company I was working for was only selling Programmable Logic Devices, PLDs. These were small, simple logic devices that customers were easily able to configure for whatever logic functions they needed. The company knew that, in order to grow, it had to introduce Complex PLDs, CPLDs. These were larger and higher integrated devices towards which many customers were moving. To give you a comparison, a PLD had between 20-28 total pins. A CPLD started at 44 pins and spanned all the way to 300+ pins. As an organization, we thought we had a reasonable idea of what had to be done, but the reality was we only knew the surface issues and not the specific details. The software support was significantly more complex, the programming details were much more detailed and so on. Fortunately, both sales and the executive management were aligned and committed to do what was necessary. In fact, our long-term survival was predicated on succeeding, and we did. But I've experienced many other new product segment introductions that were failures. The effort required was enormous, and the understanding was simply not known. The bottom line is that marketing and the entire organization must learn quickly and be prepared to invest if you want to be successful in a corporate positioning change.

Often companies are tied to their past, and this can be the death of a high-tech organization. The organization should be regularly looking at where their products stand and in what direction the market is heading. Also, don't be afraid to eat your own lunch before someone else does. This is a saying which means if you don't replace your product with something better in the future, other companies will. In an intensively competitive market, firms with a leading position should cannibalize their own current advantages with next generation advantages before competitors step in to steal their market share. The only time this is not true is when the company has some very exclusive technology, patent or IP that cannot be replicated without enormous effort from your customer to change. Let me share a history lesson to highlight the lack of strategic vision regarding corporate positioning. In the United States, just over a hundred years ago, trains were the major source of transportation. Automobiles were being introduced, and airplanes were being improved upon at a rapid pace. This was evident by many accounts at the time. Have you ever thought, why is it that not one railroad company became an airline organization or an automobile manufacturer? The reason is they were all myopic about their corporate positioning. Every one of the railroad executives thought they were in the train business instead of the transportation business. They should have asked themselves what business are we in, the transportation of people or the train business? If they had asked, there likely would be automobile or airline companies who also run train lines today.

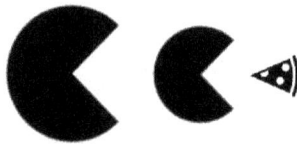

Figure 6. Corporate positioning sometimes leads to eating your own lunch

To reiterate, the product positioning should be related to the company positioning, but it does not have to completely align. If the company wants to position itself as a high-tech leader, then some attribute of its product must be a leader. Going back to the memory example, if the company makes the fastest speed or leading density, then they are backing up their high-tech leader positioning. If an organization can't execute the products, then the corporate position will not be accepted by customers.

PRICING STRATEGY

What is your product worth? Surely, opinions will vary on the answer. In fact, how to set pricing is not normally a clear-cut process. For example, if your company positioning is high-end, then the pressure to price on the upper end of the market will follow, but this is not always the case. Some companies will simply determine the cost to build a product and then add the business overhead. On top of that number is some margin number that is somewhere between the required minimum margin and a higher desired margin number. Although this is a simple math equation to calculate pricing, you should not necessarily use this answer for your pricing.

Before you jump to using the equation to determine your pricing, I suggest you first determine what you think the customer will likely pay. Look at the competitive offerings or alternatives. See how they are pricing their products. You can do this by asking customers or distributors or checking various websites. Once you have a range of the competitive pricing, you can use this as a starting point. From there, you should factor in your market share, product differentiation that customers value and brand strength of both you and the competition. Consider not just your product, but your solution's strength or weakness and how valuable these are to customers. For example, if you sell to the defense market and have a certification approval that no one else has, you can likely charge more if that is important to those users. However, if you have a competitive product, but don't have all the solution pieces that the competitors offer, you will likely have to charge less.

Don't be surprised if the market price number cannot be

accepted by the organizations finance department. In that case you can try to sell at the higher price and monitor the results. The risk is that you will be labeled as high priced and may have to work even harder once you adjust to the market pricing. Pricing can be changed at any time, but it should not be done often. It is easier to go down in price than up, but avoid pricing things so high you get a reputation for overcharging customers. Above all, don't ever gouge customers with very high prices. If supply becomes tight and you have available product, assuming no contract pricing, you can consider charging more, but don't over charge them excessively. If customers believe you are gouging them, they will remember. Customers often have very long-term memories. When supply comes back, you can be banned from the approved supplier list, designed out or even banned from contacting engineering. I've seen all these happen more than once in my career.

Some products will follow the bathtub curve of pricing. This sometimes occurs to commodity devices but is more often true for proprietary devices. By proprietary devices, we mean that there is no direct substitute or pin compatible device to replace your device. The initial proprietary device pricing is high because the product is very unique. In time as the product matures, the pricing moves lower as competition copies your solution and introduces alternatives. After several years, when competitors stop supporting their products or the offerings become scarce, you can raise prices. This is often done by companies to force customers to pay more for the older devices and encourage them to move to newer offerings.

If the product you are producing is a pure commodity, the competitor's pricing will be a major influencer of what your pricing will be. If the product and solution is a proprietary offering, then the customer value will be more of a factor. However, know that even a proprietary solution has alternatives. If the alternative solutions are close to your product, this will make your product a quasi-commodity. It may not be easily replaceable, but it can be unseated by the customer with some effort on their end. The point is, don't ignore the competitive pricing of the market. In fact, many products may be introduced as proprietary and often move to commodities or quasi-commodities. Your pricing will have to adjust as the number of competitors and the market realities change. In the next chapter we will talk about setting goals. Your pricing strategy should support the goals you have set for the product and solution while also considering the strength of your offering. Settle on pricing which supports the product goals and allows the company to make adequate return on investments.

4

SETTING YOUR GOALS

BELIEVE THAT YOU HAVE TO BRING REVENUE IN

At this point, you should have a clear idea of how you want your product to be accepted by the customers and what are the key market segments you will focus on. Now you need to set goals. As you think about what goals you will set, you need to internalize that it is up to you to bring in revenue. Feel this positive pressure as you consider the goals. Every successful product or solution I have promoted and sold, I felt personally passionate about marketing it as best I could. If you truly believe in what you are marketing and selling, your passion will be evident to sales, customers and the entire organization. Keep that in mind as you set your goals.

With a product and solution defined, the segments chosen and positioning selected, you can now move on to setting one or more goals. Don't make a mistake and just start marketing without goals. If you work for a large corporation, your goal should probably align with revenue generation. If you are in a smaller business, it must. You should develop a mindset that the survival of the business is dependent on you, and that you can be the difference that will bring in the revenue and profit for the company. At the end of the day marketing just to do marketing makes no sense. It is a waste of time. You need to set a goal or two for what you are bringing to market.

CREATE A SALES FUNNEL

Creating a sales funnel can help you set a goal. A sales funnel is often described as the ideal path that a consumer takes to become a customer. To help clarify, here is an example of a goal I had for a product campaign. We were introducing a new solution, and we wanted to achieve $25M in sales in four years for the product. This device required that engineers adopt it, integrate it with their end product, test it and then ultimately, ship it. The organization had created as much of the solution as possible to reduce the sales barrier. We knew there were going to be challenges, and it would take some time before we would achieve actual revenue. Our sales cycle typically took time to develop, and we took that into account when we set our goals. We also knew that each engineering customer, on average, would purchase $100K once they were in production. So, we knew we had to win 250 customers. ($25M sales / $100K average).

Figure 7. Typical sales funnel

How do we win 250 customers? As you can imagine, we had to show our product to a much larger number of engineers because not all of the customers were going to use our product. Our historical conversion rate was about 20%. Let me explain what I mean by conversion rate. For example, for every five customers we engaged with about our product, we would win one. So, we needed to create a campaign that worked with our sales team to promote our product to at least five times as many customers over our goal. With a goal of 250 customers, that meant we had to promote to at least 1250 customers, (250 x 5). This was our minimal goal, 1250 customer opportunities. By getting to at least this number we had a path to ultimately earn $25M. Often, if our campaign was going well, we would increase our goals further to better improve our odds of success. In the funnel diagram, the 250 customers would be in the purchase end. The 1250 customers are in the familiarity or consideration section of the funnel.

Customer opportunities are one metric that is back further in the funnel from revenue that you can use as a goal. You may want to track other metrics that are earlier in the funnel.

Perhaps your organization has experience knowing that a particular number of data sheet downloads leads to a typical number of leads or opportunities. Whatever the specific metric, consider these as marketing goals for you to select.

Depending on the sophistication of your marketing tools and experience, you may have a few goals in different places in the funnel. You may want to achieve a certain number of website visits or generate a specific number of leads or increase awareness by hitting a social media following goal. In one organization I had worked for, we created a point system for downloading particular web collateral of the device and solution. We believed when someone downloaded a data sheet and two other pieces of collateral, that this individual was a lead. We used this capability to set goals further up the funnel.

Take some time and think through what goal or goals you want to have for your marketing campaign. In your existing sales cycle, how do you convert customers? How many prospects do you need? If you are struggling with how to define a goal, imagine it is a certain amount of time in the future and you are looking at various numbers in the funnel. What does success look like? At the end of the campaign would you, your boss, or others view the result as a positive achievement? Work backwards from the sales goal to create your key marketing goals.

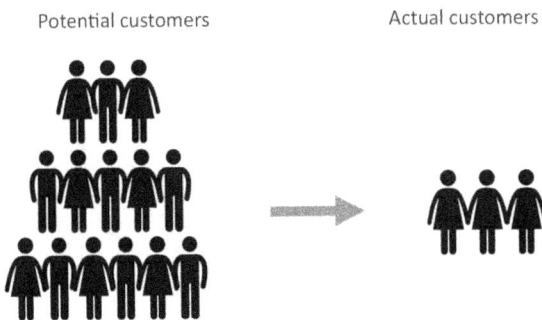

Potential customers   Actual customers

Figure 8. Setting your goal visualization

If you work in an organization that has salespeople, manufacturing representatives or distributors, think about what will help them when you set a goal. We will share more details about collaborating with sales later. Overall, when you are promoting a product or solution you generally think in terms of a funnel. Think of it this way. I have to make so many people aware of my product or service. Of those, there will be a certain number that are interested enough to become familiar. A percentage of those will consider what you are offering. A subset of these individuals will choose to purchase it. One of these milestone numbers is likely a candidate for your marketing goal. Remember your chosen market segments as you think about the goals you will create. For some solutions you may have different goals for each segment, or you may decide on a single aggregate goal or both.

ARE YOU INTRODUCING TROMBONE OIL?

A word of caution before we proceed. At this point, take a deep breath and ask yourself a tough question, "Is what I am promoting trombone oil?" I recall reading a line similar to this from Bob Iger, the former CEO of Disney.[5] The point is that, although there is a market for trombone oil, the quantity sold every year is very small. In thinking about the solution you are putting together, ask the difficult questions, and explore the market potential for your product or service. This will help you set realistic goals and appropriate expectations.

If you work in a large company, you may not have a choice to change everyone's thinking. Figuratively speaking, I once promoted trombone oil. A previous company I worked for had many business units, or BUs. There was a solution that I noticed which used devices from two BUs and the software from a third. No competitor could provide as complete a solution as we were offering. At the time, I thought we could package this and promote it as an all-in-one solution. My marketing mind was thinking that we could win a majority of these designs, because everything was within our one company and customers would find this valuable. All of this was true, but the reality is there were only a very, very small number of customers who needed the solution.

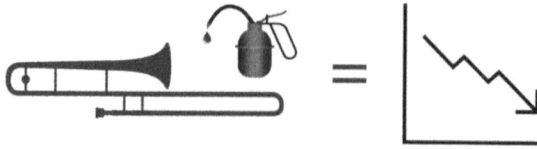

Figure 9. Avoid selling "trombone oil"

Looking back, I wish I had asked more difficult questions about this solution. I didn't know enough about the entire market opportunity, and no one spoke up to educate me on the small size. It may have been that I was afraid the answer was to not promote this at all. That is the conclusion I should have drawn. When one of the BUs only assigned a junior software engineer to the project, it should have been a huge red flag. They were just happy to have a marketing person in another BU, me, promote their technology with minimal effort from their side. I should have asked why no one had put this solution together already. A lesson was learned, and I just moved on.

To recap, set at least one goal for your campaign. Please don't skip this step and just start marketing or promoting. Without some goals against which to measure, you will have no idea what impact you are having or what changes to make to your marketing activities. Depending on how your company is organized, it may be best to have goals for each segment. With at least one goal, you can now begin to think about the different collateral you will create. Get excited about the process because the creative part of marketing now kicks into high gear as we continue with the process.

5

CREATING YOUR MESSAGING

ALL THE CONTENT WILL ECHO YOUR TOP LINE

Now that you have goals set, we can move onto messaging. Avoid moving to this step without having chosen your segments, thought about positioning and established some goals. Remember, you don't want to just market for marketing's sake. The message creation is what I consider one of the most important steps of the *High-Tech Marketing Simplified* process. Coming up with the messaging is critical and sometimes very difficult. Your message is the communication mechanism for your product or solution story. The messaging you create is critical to the success of your product. Ultimately, you should envision promoting your offering as if you were telling a story. Your messaging is the opening statement or book title, and you add more details and aspects as your promotional campaign moves forward. I can't emphasize enough how important it is to nail the messaging. Why, you ask?

More often than not, your messaging is ultimately the reason why customers will buy your product or solution. The product and solution messaging will be the focus from which all the collateral, brochures, banners, images, websites, social media, videos, etc. should flow. Think of the messaging as the top of the pyramid, and everything that you create is based upon that. All your collateral you create should echo the messaging so your story will sink into the customers. The reason an engineer or designer will choose your solution is going to be reiterated in everything you create.

Your initial message or tag line could be a single brand name or a short sentence or both. When you begin writing down ideas for the messaging, make sure to include other stakeholders. The individuals who are part of the selling process should be consulted. Your extended marketing and sales team should know why customers will select your product. This group should be the main team that works on developing the messaging. After all, this group must believe and understand why your product or solution will be chosen by the customer.

An important point to realize is that, although you are selling to a technical audience, all people buy with emotions.[6] Early in my career I rejected this type of information. When I was sales management, I'd always bring up reasons why we won or lost based on the technical capabilities we had or didn't have. The reality is that all humans use emotion to justify the logic of our purchase. I've discussed this with many engineers, and I often receive push back or disbelief. Of course, this does not mean that your product can be inferior in all technical aspects. If your solution satisfies the minimum requirements, then the customers decision will be influenced by emotions. My point for bringing this up is to remember this when you are creating your messaging. Sometimes the messaging can evoke positive emotions which will increase your odds of the product being chosen.

As you work on creating your messaging, keep in mind that you are telling a story about your solution. The messaging will also be a reflection of the company, as the product is a part of the organization. Ultimately, the messaging should not only promote the solution, but it should help the company be seen as a leader in your chosen vertical market. Companies often strive to be seen as a thought leader in the segments in which they compete. A company which is a thought leader is considered an influential authority in the market segment. Your messaging should help the company become that thought leader. The various collateral that is ultimately developed should promote the solution and bolster the company positioning.

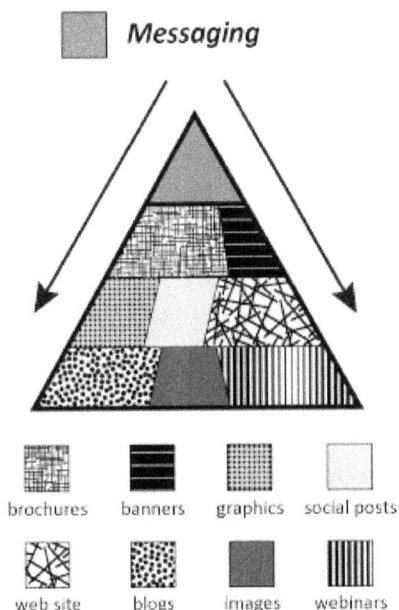

Figure 10. Messaging is the key to your story

When you work to create your messaging, you can start a number of different ways. For example, you can start by listing product benefits, product attributes and capabilities, emotions your product or solution engenders, etc. Another option is to have a more abstract message that ties to your company brand and product but is still relatable. When in doubt, my suggestion is to create a message that explains your product and why someone is going to use it or the benefits they will receive. Imagine you were in an elevator with a large potential client. The bell dings and the door is about to open, the client is about to walk out, what one line would you say about your device or solution to get their attention?

An example messaging line is what we developed earlier in my career at a FPGA company. We were introducing a new FPGA device family. The message line was, "Our devices are lowest power, cost-optimized, and mid-range density." Another example for a different product was, "Achieve higher levels of integration, security and reliability." Obviously, both of these messages explain the products features and are aimed at design engineers and technical managers. The message line for both these products is meant to educate a technical audience. These short phrases make it very clear what the key attributes and benefits of the product are. Stated differently, they answer the question why someone is going to use the product or solution.

Using product attributes is common for high tech products, but not an outright requirement. If you are marketing a technical service, you may want to point out what differentiates you from others, or what is your unique focus. A technical consultant may want to emphasize "reducing your anxiety by handling your difficult technical tasks." When you are starting to develop the messaging, write down all viable ideas. Brainstorm and don't limit yourself at the beginning. Often your messaging may come out of multiple words, lines or phrases.

Your messaging does not have to be about *what* you are offering, but *why* you are offering. Does your organization have an environmental impact goal, or is it socially responsible and giving back to the community? As an example, Tesla's mission statement is to accelerate the world's transition to sustainable energy.[7] Was there some historical importance to the business or a cause that is central to the company culture? Is there a tradition or unique technical capability that makes your company unique? Perhaps one of the segments could be used to influence your message. Consider integrating the "why you are doing business" into your messaging.

You can look to other companies in completely unrelated fields for inspiration. Maybe your product or service is associated with, or complementary to, other markets or technologies; you might be able to use pieces from those other organizations. Often there are companies that you respect or admire. What is their messaging headline? What sentence or phrase do others use in their promotions? Look all around for ideas to utilize for your messaging. Many times, it is easier to edit than to create.

Figure 11. Be creative when you are generating your messaging

Depending on how well known your company brand is and how much budget you have, you might also tailor your messaging to elicit a feeling. Although not considered a high-tech company, Subaru has simple, emotion grabbing messaging. Every commercial they make fails to mention the engine or any specifications about the car. Rather, the messaging centers around family and love. They can do this because for years everyone has known they offer four-wheel drive sedans and wagons. Their vehicles were often seen off-road. To differentiate from other manufacturers, they changed their messaging from one highlighting product specifications to one designed to elicit a warm feeling associated with love and family. This type of messaging may not be the right path for you, but be open and creative, especially at the early stages of creating your messaging.

ASK FOR OUTSIDE INPUTS

I'd also encourage you to include a wider group of people when you are working on your messaging. First, work with a few others who know your product or solution well. Individuals who know why customers will select your product. Most likely, these will be marketing colleagues or salespeople you work with. Then you should engage a wider group of individuals and ask for their ideas. It is always beneficial to receive input from a variety of sources. Be sure to get the opinions of a diverse range of individuals, including those that may not be familiar with your type of product. At this stage, engaging individuals who do not technically know your product can be helpful. Their unique perspective can generate ideas you may not have considered. If you talk to people outside the company, make sure you completely trust them or have them sign a non-disclosure agreement, NDA. Outside thoughts often provide valuable ideas.

Be vulnerable and accept constructive criticism. The best messaging is often created when many viewpoints are considered. I once had to modify a brand name of a technical device that sounded like "without, hollow or absent" in another language. We discovered this when we talked to someone who grew up in the Middle East. Although our group loved the name we picked and what it represented in English, we agreed to abandon it because our device was intended to be sold globally. You may recall in the 1970s that General Motors introduced a car called the Chevy Nova, which literally means "no go" in Spanish. How many do you think they sold in Mexico? While we are talking about being aware of names, ensure you also look up your proposed name on the urbandictionary.com site. You may not be aware of various slang terminology which your product name could represent. It could mean something with which you do not want to be associated. Use your judgement if there is a match for your brand name.

The key with your message is to keep it succinct. Making it memorable is always a plus, but most important is to drive across the key points of your brand, product or solution. Remember the main reasons why someone is going to choose your product. If you can incorporate the feelings you want associated with it, that is also an option.

Once you have your messaging, you should promote it widely inside your own company. In the Herding the Cats chapter you will see why this is important. Everyone in the organization should know your messaging headline. For example, IBM has the headline, "Unlock the Value in Your Organization with Watson." Pizza Hut says, "Nobody Out Pizzas The Hut." What you ultimately want is for all your employees to recite your message when someone asks, "Hey what is our new device or product?" The response is your messaging. The hardware and software developers who have been working on your offering will want to know, and others in the company should as well. Of course, for sales, this will be very important once you launch the campaign.

We'll discuss details later about how to create your content, but honestly your collateral will be easier to create once you have the messaging finalized. It is for this reason I emphasize the importance of messaging. Done correctly, your key attributes will shine through, and your odds of success are likely to increase. A well created message will enable customers to evaluate and adopt your solution more quickly. All your collateral will echo your headline messaging. How, where and when you deliver the various collateral is the next priority in the process.

6

DEVELOP THE GO-TO-MARKET

DON'T BE THE SALES PREVENTION DEPARTMENT

At this stage you have completed many "behind the scenes" components and now you will begin to figure out the various sales channels to deliver them to the customer. The go-to-market plan is just that. This step of the process can be how marketing differentiates your solution from the competitors. You know what segments you are going after; the positioning, goals and messaging has been flushed out. Now you will best determine how to deliver this to the customers. The go-to-market is a plan itemizing the collateral you will create and how you will drive your product message into the customer base. Overall, you want to make it as easy as possible for someone to select your device. The easier it is to choose you, the more sales you will generate. As you develop a go-to-market plan, it is important to include more than just the sales and marketing departments. Factory or headquarters-based executives throughout the organization should be aware and understand what the go-to-market plans will be. In fact, many executives outside of sales and marketing should be included in the overall plans. I would say it is critical for the broad organization to understand how the company plans to win in the chosen market segments. After all, the revenue that the company generates is what pays everyone's salaries.

Every employee should understand that they are all a part of helping sell your solution to the customers. Nothing drove me crazier than when I heard others in the organization say that the company has a sales problem. Once we were selling a device that needed software to go with it. The hardware device was quite capable, but the development software was atrocious. When customers evaluated our software, they were rarely ever able to get the device to do what it was capable of. As we lost sales, guess who was blamed? It was the sales and marketing teams that were the problem according to our executives. Despite us screaming about the software inadequacies, it seemed like no one was listening. When the company has a sales problem, it means the organization has a major problem and everyone had better pitch in to turn things around.

Figure 12. Listen to salespeople, it might save your company

When we were not supported by some in the company, we often referred to those individuals as the sales prevention department. For the sales organization to convince our company about the software issues, we arranged for the software managers to visit the customers. They heard for themselves what we were complaining about. The software managers felt the heat from the customers directly and then understood they were part of the sales problem. The sales and marketing team was screaming until we were blue in the face, but the message was only heard when the customer delivered it directly. If you have individuals who are part of the sales prevention department, get them in front of the customers who will voice their concerns.

As you think about engaging customers and your sales process, think about how you could make it easier to adopt and differentiate your solution. Both in person as well as online should be part of your go-to-market. Are you able to move more quickly because you are a smaller company? Do you have a more automated method to handle ordering or technical support or other aspects of the sales cycle? Perhaps your organization is willing to provide more in-person technical support and your competitors have moved to email and forums as their only support. These differences have little to do with the product features but could be huge differentiators for customers. Think about how you can push what you do better than what the competitors do. If you can use online tools or a mobile app to make it easier for customers to use your product, that should be leveraged.

You should also include company executives in other parts of the organizations in the go-to-market plan. It is amazing that marketing and sales can tell executives what they believe is important for the customer to hear, but often it does not register. When those same executives visit the customers and they tell them exactly what marketing and sales have been saying, it is astonishing that the message is then completely absorbed and accepted. One way or another, you want to gain the trust and respect of executive management and utilize them as part of your go-to-market. Sometimes the only way to do that is to have a direct customer to executive meeting.

As you are creating your go-to-market plan, be sure to consider what the competitors are likely to do. If you are a small player in the market, the larger competitors may not react at all if they do not see you as a threat. However, you should think about and anticipate how competitors could respond to your new solution. Not that this should be a major part of your plan, but you should anticipate how the main competitors will react to your product and solution introduction. If in your go-to-market collateral you have a slide that shows how you compare to the competitors, you'd better believe they will react and put out their own versions. Generally, I think it is best to focus the majority of the go-to-market activities on your customers. After all, they are going to make the final decision. Just realize that competitors are not likely to stand still, especially if your offering is viewed as a threat.

For semiconductor and many high-tech products, the go-to-market sales process goes beyond digital mediums of the website, how-to videos, product manuals, evaluation kits, social media, etc. It is usually higher touch and there are often both salespeople and distributors or value added resellers (VARs) that are the front lines of the sales process. The sales team is usually either direct sales individuals who work for the company or an independent manufacturer's representatives (reps). This is usually the main selling channel for your devices and solutions. Distributors can also play a vital role, but that will depend on the specifics of your product. Let's discuss the sales channels first.

SALES CHANNELS

Let me explain what a manufacturer's rep is first. I'll refer to them as reps. A rep is an independent organization which promotes tech products that are complementary. In exchange for promoting the products, they usually receive between 2-7% commission for each sale. The rate depends on the complexity of the devices, the solution and many other factors. Reps are contracted by companies to represent them in a particular geography. Reps do not normally stock products, but some may offer that option. Companies usually use reps when they do not want to or can't afford to hire their own salespeople, or they cannot support a region directly or both. Reps do not normally have competitive products, but they do offer complementary devices. All the devices that a rep promotes are called their line card. These are some of the key values that reps bring to the sales process.

Reps often sell other devices on their line card to a customer, and because they have that relationship, they can introduce your product. Well managed rep companies can be as effective as a direct sales force when handled in the appropriate way. What do I mean by the appropriate way? I have plenty of examples of what not to do, but let's focus on what are the best practices. The most effective rep relationships I have experienced was when the company treated their rep as their direct sales team. They trained them as if they were direct sales employees. The rep organization was also told if they grow customers to be very large, that they would be able to keep those customers. Often manufactures will pull a growing customer from the rep so they can service the business on a direct basis. Lastly, when there was a problem with an individual at the rep, an open discussion with the rep owner would take place about how to improve the person or how to replace them. With reps, what you invest in them will return back to you. I've been amazed how few times this lesson has been learned by semiconductor companies. Reps are often treated as second class citizens, and that is simply wrong.

Direct sales are easier to understand. These are individuals who work directly for the company. They are often responsible for a territory, a collection of customers or a vertical market segment. Just because a company has direct salespeople does not mean they are better than a rep. They may or may not be. Direct salespeople, just like reps have to be trained, invested in, nurtured and supported. Depending on your company, you may have only direct sales, manufacturer's representatives, or some of both. Many larger organizations tend to have only direct sales. Some companies will have direct sales in the larger markets and reps in smaller geographies. Regardless of what type you have; salespeople can make a huge difference. Customer relationships matter when selling technical devices. In fact, some organizations go as far as hiring manufacturing representatives who specialize in a specific customer segment. For example, the automotive, military and defense markets often have specialty supply chains, and if a rep is part of that, they can be very helpful to win at those customers.

The last, but arguably the most important, role in the sales team are field application engineers (FAEs). Normally, an FAE works for the company and is the first line of support for customers that have technical issues. FAEs must be both technical and communicate well with customers. They are often an invaluable piece of the selling cycle and should be integral to the go-to-market plan. Usually, a salesperson or rep will qualify a customer. If the customer is a good candidate, and they are interested, the FAE is next brought in to dive deeper into the technical capabilities of the device and solution. A good FAE is a valuable resource, and many times closes the sale from a technical perspective. FAEs can work well with reps or direct sales models. Occasionally a rep will hire an FAE, but normally they work for the manufacturer. These are the various sales roles found in semiconductor and high tech companies.

Now we will walk through some go-to-market examples for you. Let's say your plan is to focus only on the top 10 customers. Assume you can determine the top 10 accounts you must win. With this short list in hand, you then create a plan to earn design-ins at these accounts. (A design-in occurs when a customer selects your product and starts integrating it into their product.) This plan is led by sales but would not only include what the strategy and tactics the salespeople will use for the account, but also what requirements outside of your solution are needed by marketing or engineering to create. Marketing should identify if any company executives have a relationship between themselves and the top customers. The goal is to have one or more corporate executives in your company know a key executive or influencer in the customer. If that relationship exists, it should be cultivated and work in concert with the sales team to increase the understanding of the customer, their key requirements, upcoming product plans, etc.

For example, if your VP of manufacturing knows the VP of quality at the customer, then this connection should be leveraged. Arrange for a few meetings between these individuals so it can be determined what is most important to the customer from a quality perspective. This might change your solution and improve your odds of winning at the engineering level. Of course, this will not be the only connection at the customer. Sales and marketing should be calling on the appropriate engineering organization, as well as component engineering and supply chain groups at the customer. When you sell up and down in the organization, your odds for success are greatly enhanced. Sometimes reps have excellent relationships with specific customers. If this is the case, you should integrate them into your go-to-market plan. Sales is not just the sales team's job. Everyone can, and should, help sell. As mentioned earlier, executive management should be involved with sales and marketing to create an organizational go-to-market plan.

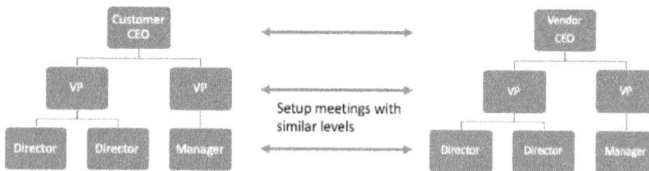

Figure 13. Arrange meetings between your customers and your organization appropriately

As mentioned previously, when factory-based individuals visit the customers in the field, there can be positive benefits. However, it is important that they are brought to only appropriate meetings. You do not want to bring in your marketing executive to meet with the customer's technician. The appropriate level at the customer is important. Also, if you want to further drive home a point you have been making to your headquarters, then set up visits where customers will reiterate what you have been saying. When your executive visits, he will hear it directly from the customer. One final point about sales and marketing groups, when factory or headquarters-based individuals visit the field, they are going to take a snapshot of what they see. If their visit was informative, productive and with the right people, they will remember that and help you more in the future. If the visit had cancelled meetings and was a waste of their time, they will also remember that. You might be in the doghouse if that happens, so be aware. Always prepare for executive or factory-based visitors by confirming appointments and ensuring everyone knows the meeting objectives ahead of time.

Your go-to-market may focus on many customers instead of a few. You may want to broadly focus on your chosen vertical market segment or several segments. In this case, you need to think about promotions and how to get your message out about your solution in a broader way. You might consider conducting onsite seminars and events where engineers can touch and feel the solution. You could also use webinars, online events and other virtual opportunities which are much more scalable. Recognize that many engineers have a mentality of "show me or prove to me your device does what you say". Understand that organizing these in-person and virtual events, the presentation, message, and positioning should be aligned with what was created for the solution. You should be able to leverage the collateral you will be creating. Ensure that the sales and technical field support people who are conducting the event have the appropriate training and clearly understand the messaging and positioning. You want to ensure that what the customer hears in person echoes what is being pushed via virtual, social media, videos, websites, etc.

As you think about how to create a go-to-market plan, make sure you include your sales channel resources and many other groups across the company. Sales and marketing resources will be your most impactful in driving the message to customers, but others in the company can help. In addition to the digital mediums, in-person sales and virtual events that you will create, there is another channel that can help you in your go-to-market: distributors.

DISTRIBUTION DETAILS

Distributors, like reps are often misunderstood. I will use the term "disti" occasionally to refer to a distributor. A distributor is a company that stocks products and sells them to customers. They often have an extremely broad line card with numerous competitors spanning an extensive array of technology. For semiconductors there are two very large distributors, Arrow and Avnet. There is also a very large private company distributor, Future Electronics. These three dominate the traditional semi-conductor distribution channel. A number of smaller and regional distributors also exist. There are also what are considered catalog distributors. The two largest being Digikey and Mouser. They used to send out printed books or catalogs of all the products they sell. Some still have print catalogs, but most do business mainly via the web now. When you put together your go- to-market details, think about how distributors could be a part of it.

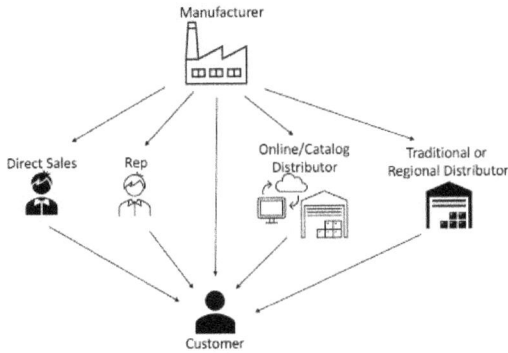

Figure 14. Manufacturer to customer sales channels

Distribution can be a key differentiator for your go-to-market. If you have one of the large distributors Arrow or Avnet, and you are a small company, you will likely not receive attention or support that you want. However, both are on approved vendor lists at a wide array of customers. Catalog distributors like Digikey or Mouser make it easy to order online and can help with direct customer promotions. There also exist regional or smaller distributors, where you might receive more attention, but just having them alone means less coverage and potential barriers for customers that don't use them. In my experience, it is best is to have one big, at least one catalog and consider another to balance out the bigger disti.

Some distis are vertically integrated, such as focusing on military and defense which can be very helpful if that is your chosen segment. Distis who receive better support and margins can help steer business toward your solution. Just as salespeople and manufacturers' reps have relationships, so do many of the distribution salespeople. Often, when looking at selecting distributors, you can ask who their top customers are. This will help you decide if their customers line up with those you think are the best targets for your solution. Some distributors have actual stores and inventory on site at larger customers. This can be a major competitive advantage and further differentiate your solution. Educate your organization's leadership to see distribution as a valuable piece of the sales cycle. They play an important role both from an inventory perspective, flexible pricing terms, low volume sales, customer base expansion as well as the support they can provide on a local basis.

I've experienced all types of emotions and opinions when it comes to semiconductor distribution. Perhaps you have heard of the frustrations about working with distributors. This one saying I heard has a lot of truth. A disti is coin operated. Meaning that when you sell via a disti you can accomplish more by providing them higher margins. All kidding aside, distributors can be an effective extension of your go-to-market plan. To include them, you need to keep a few things in mind. First, never create a grand strategic plan for a distributor. Distributors need clear, simple direction with the appropriate support to successfully sell your products or solutions. They can do things for you if you give them a reward. That reward which they want more than anything is margin, gross margin. If you want their attention, then you need to share margin with them. If you want them to help provide technical support or training for your products, then you need to provide even better margins.

Some distis specialize in commodity sales. They are all about inventory, efficiency, and short turn around on quotes and quick shipments. Other distributors are better at proprietary sales. Some are stronger in certain geographies or market segments. Knowing the strengths and weaknesses of your distributors will help you create a go-to-market plan with improved odds of success. You must create the appropriate training material, simpler collateral and other support mechanisms required for distribution. Remember, the distributor may have hundreds or several hundred lines that they sell. Because of their breadth, distis are often diffused. This is why dealing with them means you have to keep things simple. They may try to focus only on the top 15 lines, but their line cards are very broad. Whether you are one of the top lines or not, if you decide to include a disti in your go to market, then you have to align your sales team appropriately with the distributors.

If you make it easier for a disti to do business with you than your competitor, then you will do more business with them. Assuming of course that most other technical factors of the solutions are comparable. For example, I worked at two different companies that approached distribution very differently. Both companies sold products that were proprietary devices. Meaning that once the device was designed in, it was sole sourced, and the customer had to buy the product from us. One company was very rigid about how they quoted the devices. If a customer was going to buy 1000 devices a year, it was quoted a quantity of 1000. If that customer designed in the device and wanted to buy 10 pieces for a prototype, our company required a new quote and always charged much more for those 10 devices. This was frustrating for the distributor and the customer. Additional quotes, as well as new purchase orders, had to be generated. The company did this because they wanted to squeeze every dime out of customers it could. The distis quickly learned to hate this and started selling more of our competitors.

Figure 15. Make it easier to do business with you. You
will win more business

The other company I worked for would quote a quantity of 1000 pieces at a similar price. When the customer wanted to buy 10, the company allowed the same quote to be used for 10. The distributors appreciated this because they were able to take orders quickly and the customers appreciated the simplicity. This also saved time for all the companies involved. The manufacturer felt it was more important to win the customer and sacrifice that small margin. They also knew the disti would push their products more often because everyone knew how easy it was to sell. The bottom line is to make it easier to do business, so you will do more business!

Semiconductor companies, or original equipment manufactures (OEMs), sometimes have a registration program with distributors. If your company sells proprietary solutions where the devices are not truly second sourced, you should consider implementing a registration program. The basic concept is that you reward the distributor that designs in your proprietary device. What exactly constitutes a design-in often varies among manufacturers. Since the terminology varies, it is important to clearly spell out what is meant by a design-in. Some programs want the disti to identify and qualify the customer, then bring in the resources from the manufacture to help close and support the design effort. Others claim that design-in means that the disti must do everything from identify, qualify, convince, provide technical support, and close the business. Somewhere between these two is when a disti will be rewarded a registration. You need to clearly specify what you consider a design-in.

The mechanics of how this is done is that the list resale price is typically only 5 points or less of margin than the distributor book cost. This book cost is the same for all distributors in your system. The disti who has the registration will be given a cost much lower, typically around 20-30 points of margin. The idea is that the disti who does not have the registration will have to increase the list selling price, which will probably mean the customer will buy from the disti with the registration. This is the theory, and in most cases it works but there are challenges. Sometimes a different distributor who does not have the registration is preferred by the customer. This can cause a channel conflict and is normally resolved on a case-by-case basis. Some points to consider. Should your registration program register by each specific customer design or be organized by each customer account? Some programs are created where certain distributors target a specific list of accounts. This way, each disti agrees to focus on different accounts, which limits the conflicts. Some vendor programs remove accounts if they already have a very tight relationship. This is ok to do, but minimize the customers who are off limits, otherwise the distributor will not be incentivized to sell.

As mentioned earlier, you must align your sales channel, whether they be direct sales or reps with distributors. I knew a company that would reward their direct salesperson less for sales through distis. Corporate executives reasoned that the distributor did the bulk of the sales, so the salesperson deserved less. This caused the direct sales team to push customers to buy direct. The result was that neither the sales team nor the disti wanted to work together to win business. Successful distribution programs are aligned with the supplier's goals and aligned to the capabilities of the distributor. If the disti is best at selling commodities, avoid utilizing them to sell proprietary, complex solutions. You cannot be hands off with the disti channel. That is a recipe for failure. If you decide to create a go-to-market with a distributor, you must agree to have goals, regular follow ups, and reviews. Force both sales teams together in those reviews, and schedule them regularly. Use the phrase "we" in the discussion when you meet. Both the salespeople and distis need to trust each other. Praise the channel when they have done the right things, and suggest improvements and next steps for a plan of action. Lastly, make sure you keep the program simple for the distis.

One last item since we are talking about distribution. I want to strongly encourage your organization to be disciplined with their distributor inventory positions. Most significantly, do not stuff the distribution channel with overly excessive inventory. Companies who perform accounting with point of acquisition (POA) can slip into this bad habit to meet a sales number. POA is the agreed to cost between the manufacture and the distributor. When the product ships to the distributor, the manufacture immediately recognizes the revenue. Manufactures who use point of sale (POS) recognize revenue when the distributor sells their product to an end customer.[8] In my opinion, this is a significantly better reflection of the product acceptance in the market. Regardless, companies that use POA can slip into stuffing the channel to satisfy executives goals, Wall Street targets or other reasons. There are legendary stories about company executives who would make end-of-the-quarter deals with distributors just so they could make a sales number. I once sat in a dinner meeting where a distributor was asked to buy an additional $1M worth of product so the manufacture would meet their financial forecast. This becomes a problem as the next quarter's sales objectives become more difficult. Often this results in a vicious cycle that usually ends badly for the manufacturer. I strongly recommend businesses like semiconductors use POS.

SALES REALITIES

Over my career, I've experienced many sales scenarios and I wanted to share some important ones at this point. One of the best things that can happen to a company is fixing something when it goes wrong. Trust in your company that they will fix things when it goes wrong. Something will always go wrong; device bugs, incorrect documentation, late shipments, etc. When problems do show up, look at these as opportunities. When I was in sales and major issues occurred, I was always excited because it meant we were going to have more exposure to the executive level of the customer. Normally these key decision makers and influencers are difficult or impossible to reach. But when there is a problem and they get involved, you have an opportunity to shine.

Embracing problems at a customer works only if the manufacturer has a culture which values selling, customer satisfaction and winning. If the organization does not value satisfying a customer, then problems will be a disaster. You should do whatever you need to do to rally the key individuals that will support your customer when there is a problem. When there is a problem with a product it is not a sales problem, it is a company issue. Driving a mentality and culture throughout the company that customer satisfaction is important to growing sales is critical. If your organization does not have this, start figuring out how to create it.

If the manufacturer supports the customer, fixes the issue, and resolves the problem, you and the company will be held in higher esteem. The customer now knows they can trust you to fix issues when they come up. This is often the difference in winning the next design if most of the other solutions are close, even if yours is inferior. High levels of trust and long-term relationships can cover up many deficiencies, and often repels competitors who may have a better product. Think about how to reinforce the point that you will be there for the customer when things go wrong. After you solve an issue, it will help build trust. Consider incorporating problem resolution into your go-to-market. View the relationship and trust as part of your solution because it is!

Often sales and corporate management are disconnected. One time when I was in the sales team, we were given a product which should have sold much better than it did. This product was a new segment, ideal for the middle of the market. The positioning was great, and customers could see the potential immediately. Before this architecture was introduced, there were low cost, simple devices and high feature, expensive products. The mid-range product we introduced should have allowed the company to gain new customers and grow its market share. But as you can imagine, when sales and corporate management are not aligned, the results are not as good as they should be.

At the time, our management was keen on selling our higher end products and kept pushing us to move those. The issue was that our high-end solution was not as good as the entrenched competitors. The field sales team saw the customer excitement for the mid-range products. It was less expensive, and there were really no competitors, whereas the low end and high end had well established competitors. The sales team wanted documentation and support features for the mid-range, like the high end offered, but management resisted. Why should we sell the mid-range when we can sell the high end? This was the very short-sighted corporate thinking at the time. Our sales teams knew the high end had a much larger competitor already filling that space, and it was very difficult to win there. The debate and conflicts went on for numerous months and quarters. When the company finally committed to fully support the mid-range, it was too late. The competitors started talking about their future mid-range products in that space and the opening we had was shut. The bottom line is that corporate or factory management has to align with sales and if not at product launch, then very, very quickly. Salespeople are not dumb. They are learning every day from customers. Corporate management would be wise to listen to them more often.

From my sales management experience, I can also share that the majority of salespeople do not want to enter large amounts of data for tracking. It is important for the company to have good customer resource management (CRM) and data on their prospects and customers, but you should balance how much data is requested versus what is going to be used. Good salespeople want to be selling and in front of the customers. The more you tie down their time with entering information, the less they are selling. Fortunately, today there exists numerous tools available to make data entry easier. Just because it is easier does not mean the company should ask for more data. Be judicious and realistic about what the company really needs to know. Whatever you can eliminate from a salesperson's data entry, the more time they will have to sell. To reiterate, don't be the sales prevention department.

Salespeople, whether reps or direct, have different capabilities. Some are more technical and could help develop a market and sell more complex solutions. Other salespeople may rely more on relationships and bringing in the right resources to deal with key influencers at the customer. If you sell a commodity product you will want salespeople who focus on purchasing and qualification groups so you can second source products. For proprietary products and solutions, you will need a different skill set and likely more technical people. Marketing must make it as easy as possible for sales to succeed. Know what your sales teams' strengths are and balance that against your solution for the best possible odds of go-to-market success.

7

MAPPING OUT YOUR DELIVERY

THE HOW, WHERE AND WHEN

Now that you have the product and solution go-to-market completed, the next step in the process is to map out how, where, when and who will deliver the story of your product. While you are working on this step, you should already be creating the collateral for the go-to-market campaign. As a reminder, collateral is everything your customer needs in order to learn about, understand, adopt and use your product or solution. Knowing the various go-to-market ways you want to get your messaging and your story out, will dictate the exact collateral material you need to be creating. By this point, you should have a good idea of your target customers in your selected segments and what is required for your sales channels and distributors. From your market segment selection and your go-to-market, you know whether your product is for a broad audience or only a dozen companies. Obviously, the details of what you will do will be influenced by who you want to obtain as a customer and how you can differentiate your solution from the competition. Keep this in mind when you think about the different mediums you will be using.

Another key reminder for you at this stage is to continue to be open and think creatively. One important lesson which helped me in my career was understanding the definition of insanity, which is doing the same things over and over again and expecting different results. Let that sink in, because it was career changing for me. If the go-to-market plan you were doing in the past was not yielding the desired results, then stop doing it! For example, if the webinars and virtual promotions were not effective in the past, then stop doing them. Don't be afraid to try new things. If your past results were not what you wanted, then do something else. Stop following the rule of insanity. Look around at how competitors and completely unrelated companies are running their go-to-market campaigns. Sometimes the best ideas are modified from what you see elsewhere. Do some searching for inspiration and you will see it is often easier to copy and modify than to create.

Figure 16. Be open and don't follow the rule of insanity

An additional key piece of information is that your go-to-market will require repetition. It takes time for your message to be received. Accept that, for someone to absorb what you're promoting, it will require that they see or hear it multiple times. You can find all kinds of studies about repetition and how often someone needs to see or write something to remember it. In today's fast-paced world, where information is flying at us constantly, the number of repetitions for someone to remember is likely higher than most studies. Engineers are notorious for wanting to be convinced and reassured about new technologies and suppliers. Realize that multiple waves of product reinforcement will be needed. Regardless of how much convincing is required for each individual, understand you have to deliver your message multiple times and numerous different ways.

DETERMINE THE FREQUENCY

At this point, you want to figure out what is required to make your go-to-market plan a reality. Start by writing down all the different ways you can promote and support your device, product and solution. Common examples include your website, social media, mobile apps, sales training, blogs, technical classes, videos, emails, lunch and learn seminars, contributed articles, podcasts, meet-up groups, webinars, search campaigns, conferences, getting-started guides, design examples, press releases, free samples, etc. Don't be afraid to come up with out-of-the-box ideas. After all, some products are sold in very unconventional ways. Tupperware was sold by having parties at one's house. Who would have thought that would be successful? You may not have the money or resources to use all your ideas but write them down and prioritize them. Your website and web searches need to be prioritized but trying something new should be encouraged. Technology often changes how we receive information. Think creatively as you determine how to map out your go-to-market campaign.

Figure 17. Repetition and different media are key to allow
your message to sink in

Now let's talk about how to schedule and coordinate the communication of your go-to-market message. Most importantly, you want to create a marketing campaign that has a regular drum beat. By this I mean that you should plan on regularly scheduled promotions and communications. The campaigns I have spearheaded have usually kicked off with a bang and then kept going for months, similar to the way a lightning strike is followed by rolling thunder. For example, you might have a new product reveal that you put on your website and create a press release. This is the lightning and all the promotional actions and marketing activities you follow up with in the days, weeks and months afterward is the rolling thunder. Even though you may only be introducing one new product, think about all the different ways that it could be talked about. After the announcement, think about mentioning related aspects of the solution. It could be an eval board, a software update, application notes, how-to video, customer testimonial, sample or production availability, new example design, etc. Space these promotional points out, so the engineers receiving this information will be absorbing your solution repetitively.

One aspect in your delivery mapping that you should also

schedule are points demonstrating thought leadership. By thought leadership, we mean your company is being seen as a leader in the market segments on which they are focusing. Consider sharing your technical philosophy or application related knowledge that might not even mention your product. Perhaps your company has a unique manufacturing capability or simpler customer service process. Bring up points that are related in some way to the product that demonstrates the company's knowledge. What you are doing with these posts are demonstrating that your company should always be considered because they are knowledgeable in the market segment. The idea is to raise the technical profile of the company. If you can obtain thought leadership, then your products will likely be more strongly considered.

LEVERAGE THE SWIM LANES

So how do you organize and figure out all the deliverables and when to roll them out, whether they be product related or thought leadership? What I have found most helpful is a slide or two that conveys all the activities that the go-to-market campaign will encompass. The best representation I have found is to use the swim-lane chart. This is one or maybe a few pages that explains all the various activities you will do to promote your product, solution and company.

At the very top is a calendar layout that could be in weeks, months or quarters. You may only have a couple months on a page or several quarters of the year. It will depend on how many activities are planned and the granularity you want to show. Basically, what you want to do is list the calendar anchor points on the first row or two and then all the activities that you plan to do in the rows below. The top rows are pointing out when your initial announcement, or lightning, is going to strike followed by when the rolling thunder will be heard. The top row might be events or press releases, whatever is going to be the kickoff for your go-to-market promotion. If you don't have any major event, then you should put calendar fixed items at the top that are natural events to trigger marketing promotions. For example, you might list new product features, updated software enhancements, available evaluation boards, popular conferences, registration programs, new expected solutions, etc.

Marketing Plan through 1H CY 2023

Figure 18. Swim lane example for Q4 2022 – Q2 2023

The next several rows will be the categories of where and what collateral you are going to deliver. There is no rigid method for these rows. You might get very specific, such as a row for your web site, every social media channel, search campaign, sales collateral, blogs, etc. My advice is to list rows that remind you what is needed in your go-to-market, but don't make the sheet or sheets very detailed or overwhelming to understand. For example, you may want to combine all your social media and blogs together in one lane.

List the various rows for all the different mediums you plan to use. As a reminder, earlier in this chapter we listed website, social media, industry events, sales training, blogs, videos, emails, lunch and learn seminars, mobile app, contributed articles, podcasts, meetup groups, webinars, search campaigns, conferences, getting-started guides, design examples, press releases, free samples, classes, etc. You can group some of these together to simplify the chart or add your own groupings or ideas. As you review the go-to-market, make sure you have categories that cover all the various items. Ultimately you want to be confident when you look at the sheet that it clearly shows what you will be doing and when. A consistent and repeated communication is desirable in order for your customers to absorb your solution messaging.

To save you time, you can purchase and download my swim lane template power point presentation which has examples of swim lanes. This file also includes a thorough template of slides that you can use to communicate your go-to-market campaign to others, including your boss and executives. I will explain further in the "Herding the Cats" section why this is valuable. The power point presentation includes your goals and has template slides and swim lane examples. It is easy to use this as a starting point to modify for your marketing campaign. Purchase this file through PayPal. Click <u>paypal.me/hightechmarketingsimplified/7.99</u> to send me the funds and complete the transaction. The swim lane template PPT will save you time and make it easier to visualize and communicate your campaign. For only $7.99 the PowerPoint file will be emailed to your address.

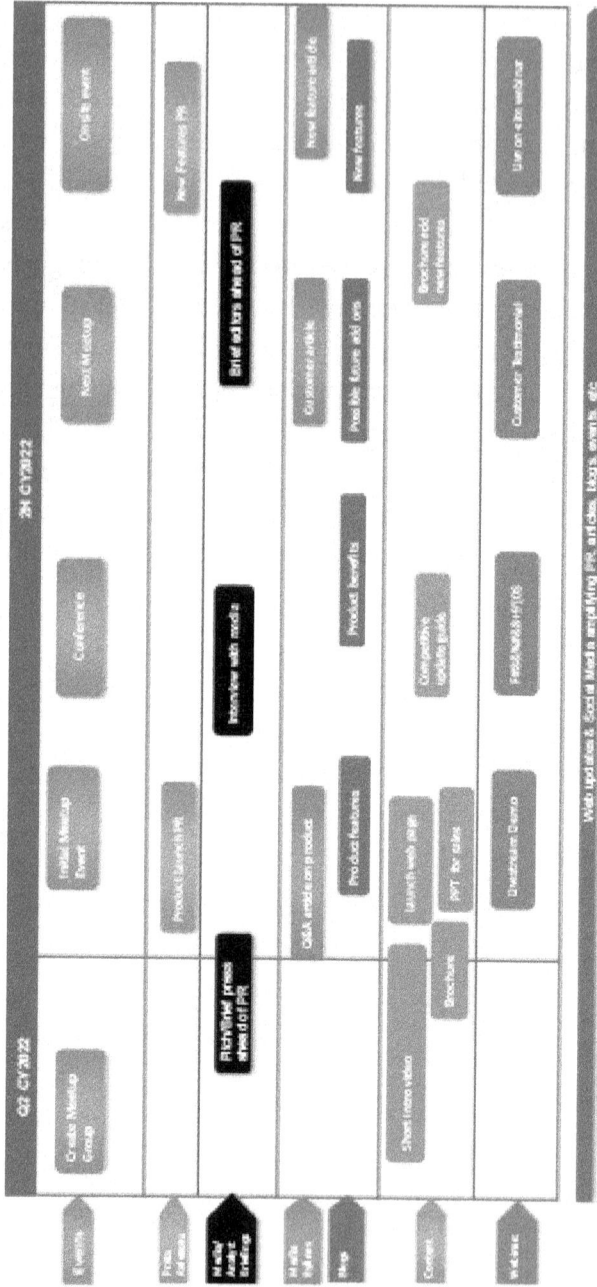

Figure 18.1 Enlarged swim lane example for Q4 2022 – Q2 2023

Avoid obsessing over the swim lane categories. What you need is the calendar at the top, then the initial lightning date, anchor events, product introductions and below that, the various medium categories. Once you have that, then you can start to fill in the various go-to-market collateral or sales materials that you will be producing. All of the different promotional tasks you are planning will each echo your messaging. When your collateral is created in the next chapter, each piece will reinforce the message of your product and solution. What you want to do is account for all the facets of the go-to-market plan. Repetition in reaching your audience with a consistent message is how you will drive your story home.

Now recall how I impressed upon you the importance of having a goal? Well, here is where it is needed to tie things together and guide the actions you will execute. Whatever your goal is, you need to make some assumptions and break things down by different marketing activities. Depending on how complex your go-to-market plan is, you could divide the goal by various sales channels as well as the different virtual mediums. You can use a simple spreadsheet to roll up to your ultimate goals. Doing this exercise will help you create a bottoms-up plan to reach your goals.

For example, let's say you have a goal of 500 leads by some specific date. You make assumptions for each of the activities. For a webinar you might assign 50, a brochure created may be 20 leads, each salesperson will bring in 10, a meetup or conference could be 35, a disti program could be 25 and so on. My suggestion is to be very conservative about assigning leads to social media posts. I often think of these activities as make up for other areas that fall short or help you exceed your goals. Remember, it takes multiple times for people to remember your product, so it would be more realistic to have fewer leads in the early campaign activities and then increase as you move forward. Your first webinar might only generate 25 leads, but the next one 3 months later might bring 50.

If you want to move up earlier in the sales funnel for your campaign, you might assume that a certain number of seminar attendees or data sheet downloads will nurture into a lead. Suppose that for every 25 downloads you think you can average 1 lead, then you could create goals for downloads as well. Don't go crazy creating more goals but do create these smaller goal assignments so you can see your progress once you get the go-to-market campaign rolling. Breaking down the ultimate goal into smaller pieces enables you to clearly see how to accomplish the task and communicate it with others.

Plan on over-subscribing to your goal once you fill in each entry. Consult with the sales team to see what they think is realistic. Often, they may not be accurate, but they are another resource to consider. Remember, your customers are busy. They are not just waiting for your go-to- market campaign. Set goals that are a stretch but achievable.

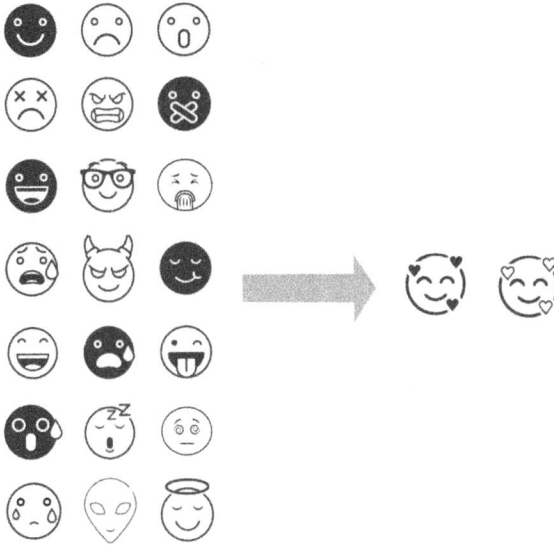

Figure 19. Reach more individuals to hit your goals

To reiterate, use multiple methods to get your message out. It will take you longer to reach your target customer than you think. Go beyond what you believe you need to do to increase your odds for success. My suggestion is to list out all the details on a swim lane page or multiple pages. The activities you think will generate more opportunities should be detailed out. Events that you are not confident about can be grouped together. Depending on the scale of what you are doing, you may not want to call out every detailed post or activity in the swim lane chart. Obtain feedback from your sales resources and a few distributors. They might have some suggestions on what works best.

THE IMPORTANCE OF SEARCH

The point of mapping your delivery is to write down all the collateral you will be creating and when, as well as in what medium, you plan on delivering it. Speaking of medium, I'd suggest you strongly lean on search engines, your website, and videos for whatever you roll out in a virtual manner. Of course, whatever you are implementing should align with your sales channel and distributor programs. Online is surely an important part of your go-to-market, but I suggest you focus that energy. Many studies that I've seen for online marketing shows social media is a smaller influencer than search engines, your website, and videos. This recent article[9] mentions 68% of buying decisions begin by searching. There is increasing belief that social media is influential in purchasing decisions, but this is focused on consumer products, not high-tech solutions. There is value to using social media for technical products, but this should not be the complete focus of your campaign. The bottom line is to ensure that search, your website, social media and videos are a key part of your virtual messaging campaign.

Another thing to consider is this: I've worked with many marketing professionals who think they cannot start a go-to-market campaign until they have a product in hand. My experience says this does not need to be the case. If your company has a hard rule of not starting promotions until the product or solution is ready and in stock, then tow the line. However, if you have flexibility, you can start promoting your product earlier. The key is to not start so early that you excite people and can't deliver within a timely manner.

If the product or solution you are delivering has a long design-in lead time or lengthy negotiation process, then you should consider starting your campaign before all the product and solution deliverables are ready. For example, we were introducing a device in a new technology that required customers to change their software infrastructure. This was not something decided overnight, and the actual conversion took months once it was started. We decided to kick off a campaign focusing on the technology, the benefits of it to the customer and so on. The sales team was trained, and presentations were created to reinforce this. We did not talk about the actual product until 6 months after the start of the campaign. This allowed us to spread awareness, start building interest and generate customer meetings to better fill our sales funnel once we actually announced the product.

Lastly, not to get overly specific, but you should always plan to put out your online messaging between Monday just before lunch time to Thursday afternoon. This is the most likely time to grab people's attention.[10] This tends to be best for the tech industry, but your product may be vastly different. From my experience, the majority of organizations put their content out during the referenced time frame. Individuals tend to look for technical information more during the Monday just before lunch to Thursday afternoon time frame. Of course, if your product is targeted for night owls, weekend use or has other unique characteristics, you should alter when to put out your content.

The last consideration in your mapping out the delivery is to create an estimated budget for all the tasks you want to do. Depending on the size of your company, you may not be able to do all the things you want. At this point you should create a budget to evaluate projected costs to deliver all the seminars, evaluation boards, web resources, videos, etc. If your marketing budget does not support all the deliverables you want, then you will have to prioritize what are the important deliverables which will yield the best return on your investment.

8

BUILDING YOUR COLLATERAL

THE CONTENT WILL ECHO YOUR MESSAGING

With your segments and positioning chosen, goals in hand with the messaging and go-to-market plan complete, as well as delivery of your story mapped out, you can finally start creating your collateral. I've always considered this part of marketing to be fun because you can work with artistic people to help create compelling material for customers to absorb. When I talk about collateral, I am referring to all types of items needed to execute your go-to-market. Collateral could be multiple graphics, a web page banner, a product brochure, images, evaluation units, videos, drawings, pictures, data sheets, web site forms, samples, how-to guides, etc. Collateral is everything that your customers need in order to learn about, understand, and ultimately adopt your product or solution. Everything that is part of your swim lane chart, and the go-to-market plan needs to be created.

Use your messaging headline and ensure that it is prominent in all you create. Notice that I said prominent and not front and center. You want to reinforce what it is that differentiates your product. What are the key reasons why an engineer is going to use your solution? Each piece of collateral does not have to feature your messaging as the headline; however everything should weave your messaging tag line somewhere in it. Let your messaging be the guide that drives the creation for your collateral. For some of your collateral you may want to emphasize the more specific technical attributes while other pieces will be higher level or focus more on the benefits. Overall, you should have the messaging woven throughout the vast majority of what you create.

For experienced marketers and established manufacturers, having your logo on the collateral is a given. Your company name and logo should be on all the collateral. If you are a startup and creating your brand logo, think about a design that is memorable. Ideally you want customers to see your logo and recognize it is your company. For most companies this takes time and resources to educate and expose your customers. The ultimate goal of a brand logo is to be instantly recognizable, like McDonald's golden arches or the Nike swoosh. Place your brand on all your collateral to increase your company's awareness in the marketplace.

Here are a few more tips when you create your collateral. Use fewer words and more images, diagrams, and pictures. Customers don't have vast amounts of spare time. You need to communicate your message as quickly as possible. You want a clean look, and less text will be important for overall attractiveness. Videos should have the same look and feel as your other content. For more technical items, you will have to use a lot of text, but work diligently to have the initial cover page convey your message with minimal text and informative images. Of course, you will leverage your tag line, company logos and other graphics to convey your story and the solution benefits. In larger organizations you likely have templates to guide you, but for smaller companies, you may not. This is an opportunity to create templates, color palettes and standardize your look and feel which will allow you to create future collateral more quickly.

LOOK AT THINGS LIKE YOU KNOW NOTHING

After you create several pieces of collateral, avoid looking at it for a few days. Focus on other work or clear your mind over the weekend and forget what you created. The next time you look at it, have a more critical eye and ask yourself, "If I knew nothing about the product or solution, would I understand this?" If you created a video, what conclusion did you arrive at after it was over? Pretend to erase your mind and forget about what you know of your product when you look at your collateral. This is an iterative process that you need to go through several times to improve the clarity of your collateral.

Figure 20. Pretend you don't know anything about the product of solution when you review your collateral

Each time you are reviewing your new collateral, try to forget about what you know regarding your product or solution. You should also pretend that you are a fan of your competitor and think about their best features. It is critical that you place yourself in a position that replicates that of your customer. Assume they don't know what your offering is. Look at what you have created with virgin eyes every time you review it. Your target engineer may know about your competitor's products, but assume they are not familiar with yours.

OBTAIN FEEDBACK FROM OTHERS

It is also advisable to show your collateral to other groups in the company, loyal customers who will provide feedback, reps, distributors who are fans of your company or individuals that you trust. Look for people who are both knowledgeable about your company and also others who are not aware about your product or solution. Let them comment and provide feedback. Be open and vulnerable to their constructive criticism. In fact, encourage them to share their thoughts. You will likely receive different opinions from those aware of your product versus those who do not know much about it. Balance the feedback to improve your collateral.

Depending on what you are creating, you could also put two or three options out on social media and have engineers and customers vote or comment. If your company has an internal employee site, you could also run a couple of ideas by a wider group. The bottom line is to consider the input from a variety of voices. When in doubt, simplify your collateral. You can always have more detailed information in other specific documents. Sidestep the urge to say everything in each piece of collateral. The key is to have your potential customers realize quickly what it is your solution offers and then have collateral for them to learn more via a variety of mediums. In the end, you will deliver your message more effectively. Remember to align any in-person materials with the online collateral. The messaging and overall story should look consistent. What an engineer hears from a salesperson should be very similar to, or the same as, what is on your website.

How much collateral you create will depend on your product

and go-to-market plan as well as the resources, the time and the budget that you have. It is strongly recommended you prioritize your web page, videos, and search collateral for online promotions. Of course, data sheets and other required technical documents that engineers expect must also be created. Since most products are discovered by search and your web site, they are an important focus. Videos are growing in popularity as well and allow you more creative freedom. Make sure these items can be seen clearly on computers and tablets as well as mobile devices. The initial introduction of what you offer must be easy for potential clients to access and comprehend.

9

HERDING THE CATS

COMMUNICATING TO THE ENTIRE ORGANIZATION

After you have started developing the collateral, but before it is complete, it is time to begin herding the cats. What do I mean by herding the cats? It is an expression I use to align management and the entire organization to aid the campaign. This is especially needed for the groups or resources who will be supporting you. Mainly, herding the cats ensures everyone who is working with you understands their roles, action items and clear, specific responsibilities. Another aspect of this phrase is having the right mentality to execute the campaign deliverables. If you have included other groups and executives as you put together your go-to-market, this should be a review and an update for everyone. If other groups in the company were not part of your plan, then look at this as a way to educate them on how the company plans to succeed with the new product.

There will be numerous tasks that have to be accomplished, and likely many people will be involved. Ideally the entire organization should understand not only the products and solution that will be sold, but also the details of how the company will sell them. It is important to educate the broader organization about the segments selected, positioning, messaging and the go-to-market. Coordination and communication of your campaign details are critical.

CREATE THE RIGHT CULTURE

As you share with everyone your go-to-market plan, take the time to ensure everyone is heard. Accept all inputs and discuss ideas openly. It will take time, but in the long run it is well worth it. In fact, you should let everyone know at the beginning of your meetings that all ideas are valued. Encourage everyone to voice their opinions. Explain that all inputs will be discussed and considered as the final details become crystalized. Individuals will only speak up if they are certain it is safe to do so. Ensure that no opinion is put down, discredited, or discarded outright. You want to do this because you need everyone to commit to the final go-to-market plan. I have often referred to this as "disagree but commit". If everyone voices their ideas and they are discussed and debated, then everyone should feel like they were heard. At the end, a decision will be made. When people feel heard, they are more likely to work together toward the common goal, even if their idea was not used. In fact, Google learned about the importance of team players feeling safe and being heard. It resulted in the best outcomes.[11] Creating the right environment where each team member is heard and respected goes a long way in having everyone work toward the same goal. Drive a positive culture with these attributes. Your organization will be a better place to work, and your results will improve.

What you do not want are people in the company to work in different directions. It is a waste of time and resources. You already have enough competitors externally; you don't need fighting inside the company. Once a decision is made in the organization, everyone must commit. Even if you disagree but you were heard, then you need to commit. If you promote a culture where people feel safe to voice their opinion, you will have more committed energy for the go-to-market deliverables. You are much more likely to have their buy-in and alignment to the goals. If you think someone is working in another direction and not committed, meet with them one-on-one and listen to their concerns. Figure out what is needed to have everyone focused on the common go-to-market deliverables.

Just as you communicate the messaging and positioning of your product throughout the organization, you need to do the same for the company positioning. Every employee should understand and support the product messaging and the company positioning. As you discuss the various opportunities for speaking publicly, think about the individuals in the organization who would be best to tap. When you are scheduling media interviews, creating videos and other interactions that customers will see, make certain those individuals are prepared, understand the messaging and are media trained. Media training is critical for individuals who will be speaking with editors. Make it crystal clear what the key points are that they should be echoing and ask them to stay focused and not discuss other topics. Let these people know it is ok not to answer every question. If an editor asks about other topics, it is ok to say, "I am not sure, but I'll follow up later." Reiterate the importance to stay on message. Companies should have several individuals to utilize for outreach. Some will be for specific technical topics related to the product and others for management or financial interactions that are company focused. These people are going to be the face of the organization, so prepare them accordingly.

The most valuable piece of information to communicate the go-to-market campaign will be the swim lane presentation of the detailed marketing roll out. If you are sharing what you plan to do with executives and others, you should create a presentation which explains the segment, positioning, goals, strategy and deliverables before jumping to the swim lane specific actions. If you have received input earlier in the process, then the material will be a review, but others who are new to the plan may want to have their voices heard. Again, receive their input and incorporate it if it adds value and can be done. If you have a boss or many bosses, you need to take the time to create a presentation to walk people through the go-to-market. As I mentioned earlier, I have created a template PowerPoint file which you can use as a starting point for this type of meeting. This includes slides for your positioning, goals, messaging details and swim lane examples to save your time. Purchase this file through PayPal. Click paypal.me/hightechmarketingsimplified/7.99 to send your payment and complete the transaction. The swim lane template PPT will save you time and make it easier to visualize and communicate your campaign. For only $7.99 the Power-Point file will be emailed to your address. I've found this type of presentation is best to align resources in the company and convince everyone to commit to the same goals.

Figure 21. Needs no explanation

My suggestion for a presentation to communicate to management and executives is to lead with your goals, and then create a strategy statement. The goals are tangible. Goals are what we want to accomplish, and executives will want to know them. You already have goals. A strategy is how you plan on doing it. Strategy statements themselves are not measurable, but they are intended to generate action and show how you will market. They explain how you will achieve your goals. A typical strategy statement example could be, "We are going to leverage our other solution to help us sell more of an existing product." Another example could be, "We will educate customers on the new product, so they are aware we now have an offering in that market." The selected segments, positioning and go-to-market specifics you have already created will reinforce the goal and strategy.

Having a few higher-level slides at the beginning will help you walk others through, in a logical fashion, what you intend to do. These introductory slides will explain why you have chosen the goals of the campaign and how you plan to achieve them using the different mediums. The product positioning and details in the go-to-market can be communicated to better explain the roll out of the campaign. The funnel diagram is often helpful to demonstrate that you have thought about what is required to ultimately achieve the goals. You could show a breakdown of a goal as an example. Once you have all of this explained, you can walk everyone through the swim lanes. If you have a piece of collateral or something that is far along in development, but still a work in progress, you can show that as an example. Just be up front that the collateral is still being iterated on and not final.

Once again, be open to feedback, suggestions and criticisms. Remember that some people may be looking at this plan for the first time, and they may need extra time to digest it. Once they understand the plan, their viewpoint may be valuable, as it will likely differ from those of other, more seasoned, audience members. The goal of sharing the plan details is to allow everyone to understand it and line up the resources necessary to make the campaign as successful as possible. You want everyone to commit to the goals, even if they disagree. If you are a small company, you might want to share your plan with complementary companies, those who do not directly compete with yours or individuals who can relate to your offerings. Distributors or reps could also be utilized and provide constructive feedback. These are some ideas of outside resources that may provide valuable feedback to improve your campaign.

If you are in a larger organization, don't be surprised if you need multiple meetings to explain your campaign details, not only to the same people, but also to other groups in the organization. This can be scary if you are not confident in what your go-to-market plan is. So, before you present your plan to a larger audience, make sure you have all your details understood. If you have done the positioning, goals, messaging, go-to-market, delivery mapping and have started work on the marketing collateral, then you will be well prepared. Look at these presentations as positive exposure for you to highlight your capabilities and skills. If you have done the work thus far, you should have confidence in what you present.

CONSTANTLY BIAS TOWARDS EXECUTION

After you have shared the campaign, you now need to execute. Focus on finishing the collateral and all the collection of media assets that you will need. Certainly, for the first several weeks of the go-to-market campaign, you need to have everything completed before you start delivering. A warning from my work experience, just before a campaign kicks off, there will always be doubters. If you've gone through all of the details in this book thus far, you should be ready to go, but beware of those who will ask you to put the brakes on. In larger organizations, don't be surprised if you have to sell harder internally than externally. There are several reasons why this might be the case. Sometimes managers are driven by fear. Other executives would rather keep the status quo and not do anything new. The reality is that you have to be mentally prepared to sell your campaign hard internally.

Inevitably, at the last minute someone will say they want something to be improved or changed. A few times I heard that my overall strategy was not good enough. Plan on incorporating specific suggestions that they think will improve the strategy but will not delay your introduction. Some people expect over-the-top marketing strategies along the lines of something created by Steve Jobs. Understand that while a suggestion might be a small improvement, it will not warrant starting over or causing a long delay. The goal at this stage is to execute. Your success depends on executing a go-to-market plan and a marketing strategy. In my experience, numerous strategies are going to have more or less the same effectiveness. The key is to execute. Those who execute have a much higher chance of success.

Figure 22. Execution is most important and often scares
many executives

My point is that execution is significantly more important than a grandiose idea or strategy. I once read that execution eats strategy for lunch. Most people and companies fail, not because of their ideas or strategies, but because they did not execute. Always, always bias towards execution and finishing the job. People who pontificate and say you must keep reworking a plan or take more surveys or whatever are not being realistic. If you have reviewed your campaign plan with multiple people, and your management agrees, then you should go forward. Recognize that sometimes managers and people have hidden agendas they are worried about. What if this fails? What if customers don't understand and so on. The best thing you can do is communicate the plan with the PowerPoint deck and obtain the minimal organizational alignment you need. During the presentations if you heard the inputs and considered the feedback, then you should have all the information to finalize your go to market. If you have done this work, avoid contemplating your navel. Focus on finishing the collateral and delivering and executing your go to market campaign.

10

DELIVERING

REPETITION IS KEY

Now that you have everything ready to go, the excitement and stress levels will build. It is normal for this to be the case when there are deadlines. Embrace the fact that a firm deadline, like a date, actually pushes people to execute. The key here is to knock off each of the initial deliverables. Focus on aligning what you need to get done for the launch date. Prioritize what you need to do and help set expectations for others who are working to create various collateral. Important early deliverables will likely include the web site, completing the press release, search criteria, FAQs, social posts, data sheets, product flyers, how-to guides, videos, etc. All this should be in your swim lane plan. Use it as a guide to ensure all the go-to-market details you need to deliver are completed.

LISTEN TO SALES AND ADJUST IF NEEDED

As the go-to-market campaign gets under way, make sure your digital marketing resources, sales channels and distributors are getting the word out. The marketing team should visit the field and customers to further push the go-to-market deliverables. Marketing individuals should demonstrate for sales, reps and distributors what are the key aspects to drive home when visiting customers. After a reasonable period of time, like 3-6 months, assess where things are and monitor if there are major unforeseen issues. I once sold a solution which had a major software problem. The customer would fix the device pins in the software so they could build a hardware board. The pin fixing constricted what the software could do but was necessary. As a sales manager I mentioned to the factory we need improved software for pin fixing, but they brushed me off saying that salespeople complain and don't know how to sell.

As I mentioned earlier, I forced the software engineers to visit customers in the field. I was always amazed that our developers did not know how customers used the products and solutions. When the software engineer visited the customer, he got a load of reasons why fixing pins was not only important but critical. With this direct feedback, our software was improved, and we were able to continue the go-to-market selling.

Keep in mind that, as you start your campaign, you will have to be patient. Remember that people need to see your message multiple times, maybe 6, 7 or more times before it sinks in. Therefore, I always recommend your plan to have an initial lightning event followed by multiple rolling thunders. The initial press release, website page or event kicks everything off. This is the lightning and starting point. The rolling thunder is your consistent reinforcement and replaying of your messaging such as follow up social posts, more detailed documents, additional videos, articles, ad placement, etc. Regularly delivering collateral that is in your swim lane plan will help your message sink in.

Figure 23. Repetition of your message is required

One area I'd like to bring up has to do with collateral that are designed for salespeople or distributors to use with customers. For example, if you have a demo unit or evaluation kit or something similar, it is critical that you package the solution correctly. What do I mean by that? Well, I'm drawing on my own experience when I was a field support engineer and a sales manager. It was always amazing to me when marketing delivered to us collateral that was overly complicated, poorly documented, or simply not polished enough for customer consumption.

CAN SALES USE WHAT YOU DELIVER?

If you create something for salespeople, recognize a few things. Most sales forces don't typically have all the resources that a headquarters' location has. Don't assume your sales force has time to read a detailed guide, and remember they are likely not as up to speed on your product launch as you are. Given this status, what do you need to ensure what you deliver will be a success? There are two key items you must address.

First, make sure you send a complete package. Assuming a sales office has anything other than the basics is unlikely. One time we had our marketing team create a demo kit. They sent us an accessory board with instructions about how we had to use another hardware board that was sent 6 months before. Needless to say, we never actually did that demo. The hardware board we received months before was left with a customer who was evaluating it. We were not going to remove that board because we wanted to win that opportunity. Most other sales offices did not show the demo either. When I explained this dilemma to a marketing person, he was shocked. They put lots of work into the demo, but sales rarely deliver it to customers. The point is that you should send everything needed for the demo wherever possible.

Secondly, train your sales force. It is strongly suggested that you do multiple training sessions. I'm not just talking about holding two webinar sessions. Just like your customers need repetition to absorb your message, so do your salespeople. You should do a live webinar-type training, and then travel to the field shortly after and do it in person. Better yet, have the salesperson set up customer visits and do the demo then. Have each salesperson first run the demo for you in a closed environment. Then encourage them to do the demo for your target customers. It is critical to have your sales force confident and comfortable.

If you create something for distributors, then it needs to be extremely simple as mentioned previously. Just as you trained your sales team to demo a solution, you must do the same for the disti. Whether for a distributor, rep, or direct sales, it is always best for marketing to visit the customer and show the way. If marketing demonstrates and is passionate about the collateral at the customer, then the extended sales team will likely be as well.

To recap, your campaign should have all the key deliverables ready by the launch date, and remember that repetition will be needed for your message to sink in. If you cannot create all the deliverables on time, prioritize the most important items and just keep going. Execution is most important. Not everything may go according to plan but focus and stay on schedule as much as possible. Moving the ball forward and continuing to deliver to your swim lane items will increase your odds of delivering a successful go-to-market campaign.

11

MEASURING

WHAT ISN'T MEASURED ISN'T MANAGED

How do you know how your campaign is progressing? You measure it! I once heard that in business, what is not measured, is not managed. That is pretty spot on. Whatever go-to-market campaign you implement should have some method of measurement. If you recall, one of the early steps in the process is to create at least one goal. It is at the measuring step where you can start to see how your campaign is performing. Without any goals you would just be doing things with no idea if there is an impact or if you are wasting your time. Don't go crazy measuring but do it periodically.

Whether you have decided to look at leads, web hits, page views, opportunities, downloads, visitors to your booth, customer visits, etc. make sure these are early indicators related to your ultimate goals. The goals you have earlier in the funnel stage will be referenced now. How often you review your goals is up to your judgement. It is common to review on a weekly, bi-weekly or monthly frequency, but this is based on campaigns

I've run. The bottom line is, make sure you measure and check your campaign progress regularly. After all, a successful campaign might get you a raise or it should, at a minimum, raise your profile in the organization.

Measuring is also important because you can begin to see what kind of return you will earn on your marketing investment. You may be a small business and operating with very scarce resources. By measuring your campaign, you will see what the best mediums are. This allows you to make intelligent decisions if you must cut some aspects of the campaign. In larger organizations, you will want to know what the return on your marketing budget investment is yielding. The insights from measuring also help guide what you should focus on for future campaigns.

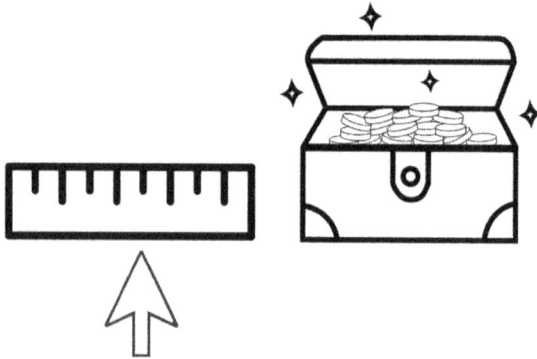

Figure 24. What isn't measured, isn't managed

You need to make sure you are following up with all your leads and potential customers. I have seen salespeople who seem to be unaware of the need to follow up on qualified leads. Don't assume that the loop is being closed. You should also look to measure your conversion rate. How many prospects are turning into opportunities and ultimately converting to wins? A lead nurturing campaign we were running was able to attach points for the items that a prospect downloaded or for pages viewed. We used web cookies to identify each person. The more an individual explored our product, the higher number of points were being accumulated. When the person passed a threshold of points, we converted them into a lead. We set quite a high number before we converted the prospect into a lead. We purposely did this, so sales would see a very high percentage of the leads being high quality.

Do what you can to have the leads be of higher quality initially. It is important that sales see the leads as having value. Once they are convinced, they are more likely to follow up. Taking these actions with sales will ensure the hard work you have put into messaging and the campaign will yield a more positive end result. Make sure you measure how your sales funnel is progressing on a regular basis. Knowing this conversion ratio will help you create better goals and focus your campaign for planning future marketing activities.

NOBODY BATS 1000%

Be advised that not every campaign is successful. The analogy can be made to baseball. No batter ever bats 1000%. Or in football, no quarterback ever completes 100% of his passes. Sometimes your product may not be right for the market or the times. Maybe your product is great, but the campaign is not hitting the mark. Or maybe you determine you are selling trombone oil. I've experienced poorly trained salespeople who are not able to close effectively. There could be a number of reasons. My advice is don't make drastic changes in the short term.

Remember your message takes time to sink in. After you have measured your campaign numbers a few times, determine how far off your goal you are. If you are within a small percentage, then maybe you just need to make some tweaks, like updates to the content or delivery medium. Another option would be to focus more energy on some of the swim lane actions that are performing better and boost them up to make up for the other activities. Early in the campaign you must resist the urge to make major changes. Stay the course and keep the effort going.

Again, let the campaign play out for an extended period. How much time you will need is going to depend on several aspects, such as how expensive, complex, or easy to adopt your product is, as well as a host of other factors. The length of time should also depend on what the normal decision time is for your product or service. If you are selling a complex product or service which normally requires six months to close, then you should let your campaign play out for at least four months or more. At the end of this period, you will either have an acceptable number of leads and should keep the campaign going, or you will need to make changes. Continue to measure how things proceed. If, after longer periods of time, the progress towards your goal is not adequate, then you will need to make some changes.

Check in with your sales team and distributors. From your own customer visits, you should begin to have a gut feel. Remember, you need to let your message sink in. Look at the early indicators in your sales funnel. How far off you are will help dictate if you should make changes. If, after an extended period of time, it is clear you are far off your goals, then you should make changes. Remember the definition of insanity!

What you change will vary. Sometimes you only need to change your expectations. Maybe the goals you set were unrealistic. Perhaps they need to be scaled down. Or it could be that more time is needed to reach your results. Of course, it could be that your content is not being received because it is too technical or unclear. Reach out to salespeople, potential customers, and leads that declined to convert to see if you can determine what are the top issues. This should help you determine if it is the product or solution, your collateral, messaging, trombone oil or something else that needs to be modified.

SUMMARY

If you have followed the High-Tech Marketing Simplified process I've outlined, you will have improved the odds your go-to-market promotion will be a success. Writing this has been both gratifying and a struggle, as I've spent countless hours translating in my mind what actions I took, why I did them and how to best explain what you should do. I'm convinced the High-Tech Marketing Simplified process can be of great assistance to many businesses and individuals in the semiconductor and technology space. It is my pleasure to have invested my personal time to share this knowledge with you. Please post a positive review if you found any of this information useful. If you have any suggestions or comments, make sure you phrase them in the form of a compliment! LOL. Just kidding, please drop me an email with your feedback and comments if you want at marketingsimplified.book@gmail.com I'm also open to book signings or speak to your group or organization. If you liked my book, please let others know. Good luck to you on marketing your technical product.

REFERENCES

1. https://www.businessinsider.com/instagram-employees-and-investors-2012-4?op=1
2. https://www.azquotes.com/author/24668-Nick_Saban
3. https://www.kirkusreviews.com/book-reviews/a/william-h-davidow-2/marketing-high-technology-an-insiders-view/
4. https://www.gatekeeperhq.com/blog/segmenting-your-vendors-to-reduce-supply-risk
5. https://www.inc.com/jeff-haden/it-only-took-disneys-bob-iger-1-sentence-to-give-best-advice-youll-hear-today.html
6. https://customerthink.com/neuroscience-confirms-we-buy-on-emotion-justify-with-logic-yet-we-sell-to-mr-rational-ignore-mr-intuitive/
7. https://www.tesla.com/about
8. https://era.org/wp-content/uploads/2020/10/REPR-F2020-ISSUE-LINKED.pdf
9. How Digital Marketing Affects Purchasing Decisions – Adlibweb
10. https://sproutsocial.com/insights/best-times-to-post-on-social-media/#find-times
11. https://www.nytimes.com/2016/02/28/magazine/what-google-learned-from-its-quest-to-build-the-perfect-team.html

GLOSSARY OF TERMINOLOGY

Arduino, Raspberry Pi – Both of these are examples of low cost SoC based boards that are part of complete solutions, including software, peripheral boards, design examples, etc.

BU – Business Unit. An organization or department inside a company typically focused on specific products or market segments. They are responsible for revenue and costs.

CPLD – Complex Programmable Logic Device. A small or mid-density logic device that is programmable and reconfigurable. Smaller in capability than FPGAs.

CPU – Central Processing Unit. A device which runs software commands to execute tasks.

CRM – Customer Relationship Management. Software and strategies that allow a company to organize and efficiently manage their customers.

Design In – When a customer makes a decision to use your device and incorporates it into his design.

Device – A silicon chip or other semiconductor device.

Disti – Distribution. A distributor is an organization which stocks and resells components, devices, products and solutions to customers.

FAE – Field Application Engineer. A technical individual who supports customers with technical assistance of a device, product or solution.

FPGA – Field Programmable Gate Array. A device that can be programmed and reconfigured to implement different types of logic or processing functions.

I/O – Input / Output. The connections to a device that allow it to communicate with another device or peripheral.

IoT – Internet of Things. This term represents the variety of products or machines which have internet connectivity.

IP – Intellectual Property. The software, hardware or other knowledge which a company or person have which has value.

Line card – The collection of companies and products that a rep or disti is franchised to sell.

NDA – Non Disclosure Agreement. A legal document companies often have other companies or individuals sign so they do not reveal confidential information or confidential data.

OEM – Original Equipment Manufacturer. A company that builds complete products by assembling and integrating many parts, devices, products, software and such.

PLD – Programmable Logic Device. A small density device that can be programmed.

POA – Point of Acquisition. An accounting term which recognizes revenue for a company when it ships it's devices, products or solutions to a distributor or customer.

POS – Point of Sale. An accounting term which recognizes revenue for a company only when it's products, devices or solutions are shipped to a customer. The shipment may be direct from the company or via a distributor that the company uses.

Product – In this book can refer to a device or a device with software or other devices.

Reps – Manufacturer's Representatives. These are organizations hired by companies that do not want to hire their own direct sales force. Reps typically represent many companies but do not have competing products on their line card.

RISC-V – Reduced Instruction Set Computing, 5th generation. This is an organization and specification for the instructions that a CPU core will run if it is to be called a RISC-V processor.

SoC – System on a Chip. This is a complex device that includes a CPU and other peripheral elements which provides a more integrated solution.

Solution – A collection of deliverables which includes a device (or product) and many other pieces of content to make adoption quicker and easier for the engineer. A solution could consist of the device, programming software, an evaluation board, data sheets, application notes, design examples, how to videos, etc.

VAR – Value Added Reseller. A company that sells a product to customers. Often, they provide additional capabilities which could include additional software, support or other services.

APPENDIX ON GROSS MARGIN CALCULATION

This section will further explain the gross margin calculation for a sale. Early in my career I made the mistake of calculating the margin, so I wanted to help explain it here. If you want to sell a device whose cost is $8 at a 20 point margin, what is the resale? Do you think the answer is $9.60? Which is $8 x .2 = $1.60 and then add $1.60 + $8 and you get $9.60. This is not correct. I made the mistake a couple times in my career and I've had distributors drill into me the right way to do the calculation.

Gross margin is calculated by dividing the cost by 1 minus the margin. Below is the formula.

Resale price = Cost / (1 − margin)

If we use the example above.
Resale price = $8 / (1 - .2) The .2 is 20% margin
Resale price = $8 / .8 = $10

The correct resale for a 20 point margin sale of an $8 cost is $10.

A MESSAGE NOT RELATED TO MARKETING

This section is not related to my book, but I feel compelled to share some thoughts. I have lived long enough to have absorbed some wisdom from a breadth of experiences. From having traveled to numerous continents, having dealt with multiple individuals of different backgrounds and cultures to having lived in different parts of the United States, my eyes have been opened. As I have matured, I've mentally fought to always give people the benefit of the doubt and look at things from the other person's perspective as best I can. Years ago, I had a neighbor who was 88 when we met. Her name was Hannah. She was the most positive, humorous, and grateful elder individual I'd ever met. Before I met Hannah, every older person I had met was always complaining, irritated, indifferent or disengaged. Hannah inspired me to see that age was just a number, and that you can be positive, balanced, and thankful your entire life.

It is my wish that everyone in the world could get together to have dinner at a table. If we all got to learn about each other over a meal, we would come to realize what I've learned. We have much more in common than that which separates us. Invite your neighbors, coworkers, and others over and agree to talk about various topics. Set the stage by saying everyone's opinion is valid, promise not to raise voices or get emotional during the discussion. Whatever the topic is, always begin at a high level where you can achieve the most agreement. For example, if you want to talk about green energy and the environment, don't run into a corner or state a position. Start first at a higher level. Ask if everyone wants to breath fresh air and drink clean water. I can't imagine anyone saying no. If everyone keeps this in mind as the key point of the discussion, then you can move down a level and start talking about how to do that. In my opinion, this is the way we need to begin negotiations and discussions for a better society. You will not always agree on every issue, and that is ok. What is important is to share some common starting point and hear what others think and believe.

Today we have access to an overwhelming amount of information. As we desire to know what is going on in the world, there are literally hundreds of choices of news outlets, publishing sites, opinion pages, and so on. The vast majority of these sites push and promote negative news, and they position the story to back up their viewpoint or political leaning. You do not have to buy in to any of them. Some people turn to social media for their information. This is a huge mistake, in my opinion. Social sites were meant to share information and images with your friends and family members or to promote products and services. They should never be looked at or trusted for news. I'd suggest you go to a number of real news sites to get an idea of what is going on. Visit both liberal, left-leaning and conservative, right-leaning sites to formulate what is news. Make your own decisions based on a few different sites. If a site continually pushes negative information, stop visiting it. You have the ultimate power to simply walk away.

As you think about various issues and topics, don't think about what politicians or talking heads are saying. Hear several sides of the discussion. Use your own experience and knowledge to form an opinion. Don't be so quick to jump to whatever side of the political aisle that you lean towards. Today we should all be thinking more broadly about how to solve challenges, not simply complain. I firmly believe we should be using AND instead of OR when considering alternatives. Consider the issue regarding utility companies wanting to penalize residential rooftop solar. The argument is that it is not fair for people who cannot afford solar. In reality, the issue is that utilities are paying for unused power generated by residential solar when the grid does not need it. The solution should not be the utilities OR residential solar should win. My suggestion is to use AND. Incentivize residential solar customers to store energy in batteries AND the utilities could draw on those batteries when the grid requires it. This enables the utilities to forgo building new power plants, which saves them money AND the solar customers are able to continue recouping the investment in their system. Everyone receives a more reliable, distributed and cleaner source of energy. Think AND instead of OR when you look to solve problems.

While I am at it, please fight the urge to complain about a problem unless you have a proposal or solution to fix it. It may feel good to whine and complain about topics but at the end of the day all you are doing is giving that subject more energy and draining your own. Think to yourself, if I were king for a day, how would I solve this problem? Complaining about things is easy to do, solving them requires an entirely different effort level. Next time you want to bad mouth something on social media, make sure you include a suggested solution to the issue.

The last piece of wisdom I'd like to share is related to jealousy and envy. Along with hate, these are very bad traits which everyone should resist. I've often seen individuals bitching about others because of what they have. Many times, this viewpoint is because the individual has not been able to accomplish something in their own life. Instead of taking personal responsibility, they become envious or jealous at what others have, so it justifies where they are in their own life. Instead of complaining about others' assets or situation, be happy for them and focus on what you need to do to achieve your personal goals. Being joyful for others is a positive attitude for yourself. After all, I'm sure you would want others to be happy for all the things you have earned in your life.

ABOUT THE AUTHOR OF HIGH-TECH MARKETING SIMPLIFIED

This is a summary of my personal and professional experience. I've reprinted it here from my previous book, *Marketing Simplified, An experience based, step-by-step guide to grow sales.*

It began in school when others just started to treat me poorly simply because they were cruel, didn't know any better, were insecure themselves or for some other reason. Perhaps it was because I wore glasses, was overweight, awkward or simply did not "fit in." Whether it was verbal or physical abuse, it all took its toll. Whatever the reason, the little confidence I had was crushed by the time I became a senior in high school. Looking back, most people experience some type of difficulty socially in school. The individuals who don't are usually those who peak when they are a senior and the rest of their life is a disappointment. That wasn't going to happen to me.

One day during that senior school year, I decided to go into the library and look at self help books. I sat down and started reading "You Can Become the Person You Want to Be," by Robert Schuller. It was the first time I'd ever had someone tell me I could do it. "It" was whatever my dreams were. For the first time in my life, I also heard my inner voice speaking words of encouragement directly to me. This new knowledge became the trajectory to turn my confidence around. I'd rely on this inner voice to pick me up and get me through tough times. "You can do it" was a regular refrain that I told myself over and over again. Little by little my confidence grew.

There was another motivator that drove me to improve my grades in school. This was independent of the inner voice. The motivation was smelly carbide dust and sweaty, monotonous, tough physical labor punctuated by occasional metal splinters. My father's machine shop, in which I'd spent my summers working from the age of 11 or 12 became a driving force, compelling me towards a college degree. When I was a junior, my high school advisor told me that if I didn't raise my C grades, I wouldn't gain entrance into a good university. The motivation to avoid working in a machine shop, combined with my inner voice started me on an upward trajectory.

That inner voice would sometimes make up contests to keep me going. In college, I felt a strong sense of competition with the other students, many of whom I believed had advantages that I lacked. In order to increase my own motivation, I pictured myself partaking in an "Educational Olympics" of sorts with the hopes of America resting on my shoulders alone. I heard that I must go the extra mile to defend the USA and be the most successful I can be. Crazy, silly, ridiculous and kind of unbelievable, but this inner voice enabled this B student in high school to graduate with a 3.6 GPA Magna Cum Laude electrical engineering degree. A few years later I earned a 3.6 GPA MBA in marketing.

Because none of my family was in a high-tech field, I had to forge my own career path. Despite the scores of people telling me there would be many job offers lined up for me when I got out of school, this was not the case. Fortunately, I did receive an offer, and my work career was underway. But all along the way, there were doubters. People let me know that this could not be done or that was impossible to do, or I was not good enough to do something. Despite the ups and down, I forged ahead. Once, to convince a hiring manager to interview me, I sent a flower basket with a note simply saying, "I know I can do the job." I did indeed earn that job, but the doubters never seemed to go away.

To grow professionally required a physical relocation for the family and myself. Nobody ever thought I would go, but I did go. Once we settled in, I noticed many interesting behaviors displayed by others. Working in many different organizations, over several years, I started learning the best practices to implement not just in business, sales and marketing aspects but also dealing with people and how to get the most out of a team. I also saw what not to do and how some things are really hard to stop doing. All along the way, the doubters persisted. All I did was to continue to focus, execute, work smarter, harder and push ahead.

Let me share some specifics of my work history, so you have an idea of how my experience led to the knowledge which I'm sharing. I actually started my career as an electrical engineer. After designing several electronic hardware boards, I determined that job was not for me long term. Don't get me wrong, I liked it for a few years, but after doing multiple hardware designs, it was already becoming monotonous. I'm not saying that engineering positions are bad, in fact they are great for people who thrive on technical details, predictable iterations and focus. It was just not for me.

After talking to many people and a former co-worker, I started in a position as a field application engineer (FAE). A former mentor who I worked with transitioned to a FAE position, and he recommended I do the same. He said my personality, combined with my technical knowledge, would be a good fit. That role was one I liked, and I performed it well for several years. This was my first role interfacing directly to customers, and it really launched me in the direction in which I ultimately ended.

I became more interested in knowing more about customer interactions and business deals. It led me to want to go into sales and sales management. I'd been working side by side with sales for several years as an FAE, and I knew I could do a sales management role. So the company I was working for allowed me to become a sales manager. At first, my territory was small, but as I grew the business, my responsibilities and area grew along with me. At the peak of my sales director roles, I was running a $100 million region encompassing about ⅓ of North America.

I enjoyed sales for the most part and I was good at it, but I noticed my influence on the future of the organization was limited. Even though I had a large territory for the company, I was growing frustrated that key customer feedback was never getting incorporated into future products. Often I was told only customers in my markets needed particular features or capabilities. I knew this was not the case, it was just that I was speaking up, where other sales people were just selling. This led me to go into marketing and business development roles. It was the best way I could influence the larger organization.

My first marketing role was focused on business development. I was responsible for working with complementary companies who could use our products. It was great to explore where the roadmap of these other companies was heading and to see how our technology could improve their solution. This collaboration with other companies helped increase our ability to track applications through increased visibility. It was also in this role where I identified an un-intended solution for an existing product. By working with my engineering partner, we opened an entirely new market application for this product. For this discovery we earned a patent, an accomplishment of which I am very proud.

After my success in business development, I was asked to take on a strategic marketing role. It was in this position where I helped define our product roadmap. At this time, the company was struggling and wanted to modify future products offerings. I worked with a product planning engineer and a product marketing manager to define a new type of device family to expand the available markets for the company. These products continue to contribute tens of millions of dollars of revenue yearly.

This exposure to all the elements of product marketing, really motivated me. The role allowed me to interface with sales, which I really enjoyed. In addition, I enjoyed teaching the sales force how to sell and promote our products. I was fortunate to work with some great marketing directors and managers. My mind was like a sponge soaking up ideas and strategies. Because I had been the recipient of marketing collateral in my sales roles, I had a unique perspective that no other marketing co-worker had. I often leveraged my sales experience to improve on every marketing program and campaign.

All of these roles helped shape my messaging, goals, campaign ideas, etc. which enabled me to create and execute compelling marketing campaigns. The time required to synthesize all the actions necessary for a successful marketing campaign were not easy to simplify. After many months and years, I was able to finish! It has been very rewarding for me to share all of this experience in a concise manner with you.

SPECIAL ACKNOWLEDGMENT
AND THANK YOU

There are numerous individuals I have come across in my life that I want to thank. I will be using first names or nicknames to protect the innocent, hahahaha. First, let me start with my family. To my wife, Janet, thank you for being understanding, caring and supportive. There have been times when I did not appreciate the important things in life, but you were always there for me. I love being your partner. To our boys, Enzo and Dante, thank you for providing us a crazy roller coaster ride of life. Continue to find your passion and pursue your happiness. We love you both and we know you are capable of great things; we believe in you. To my mother, who is no longer with us, but who I know would be beaming that I finished my second book. My father who also passed recently would have been proud of this book. Thank you to Fran who has been supportive, patient and understanding with us all. Hugs and kisses to my sisters Sandra, Diana and Maria.

Other thanks belong to my friends who know me in a unique way, Johnny Boom and Janis. Thanks for being in our lives and laughing often with us. To Chris and Nancy the demolisher, I appreciate you reminding me of my 410 score on a regular basis.

I have many work-related friends, and if you are not mentioned, it does not mean I don't appreciate you, but these were the individuals who most shaped me. I'll go in roughly chronological order. A big thanks to all of the following: Hal who taught me how to sell to large corporations; Caren for being so kind and introducing me to Janet; George X, who taught me that distribution has all the characteristics of a dog except loyalty; Snowy, (aka Steve D.) although your sales strategy of just wooing them never worked, I did learn many things about personalities, relationships and sales techniques; Cliffy for introducing me to the concept of taking forever to make a decision; Roger who explained how distribution actually works; Sean R, who was one of my top bosses, thanks for believing in me and allowing me to flourish in my first marketing/business development role; Gordon who explained to me just how many details there were to tackle in a successful marketing campaign.

In addition, thanks are due to: Bruce F who shared the magic of messaging with me; it was like a light bulb went on for me when you showed me that, and I always appreciated the collaboration with you; Sparkles who educated me on back-end web tools, lead nurturing and insights that are available. Grant J. Glad for the collaboration and laughs enjoyed while earning the patent. To Sean R who was the first great boss in my career. Thank you for believing in me. Special thanks to: Badman and Joy; we were the three amigos. I learned just how powerful synergy between complementary people can be; Diane T who taught me that nonverbal cues were extremely important to notice and understand; Timmy Tom, for educating me about how to introduce new features into future products and challenging my marketing ideas; the other Ted, who complimented me on my marketing capabilities and introduced me to new technologies; Shak Daddy, who had faith in me and often had to use capital to defend my actions; brothers, Prem, Aniket and Manu, for accepting me into the Indian harem and explaining the culture, which made me a better person. Let's get together soon and laugh our butts off.

And finally thank you to: CC who was one of my top bosses. You always believed in me and fought for me. It was also great how we complemented one another; Z for coming around to trust me and my capabilities; Loreta and David for helping turn my ideas into great videos; Erik, Ronni, Jason, Elle and Kris for sharing the secrets of social media, blogs, web and communications with me. Amber for proofreading and providing helpful changes to make this the best book possible.

If you are not listed but think you should be, email me at marketingsimplified.book@gmail.com

"Ted understands and captures a fundamental truth - that great products alone are not enough to ensure success. As he so clearly lays out in High Tech Marketing Simplified, thoughtful, disciplined marketing is the key ingredient that transforms your great product into a successful product. He demonstrates that marketing isn't magic, but it does require market knowledge, direct customer feedback and a savvy understanding of your organization's capabilities. Ted lays out a blueprint that any product marketing manager can follow. High-Tech Marketing Simplified is a must-read for any technology professional looking to elevate themselves and their products."

  - Sean Riley. CEO & Co-Founder, Legitipix

"In High-Tech Marketing Simplified, the marketing process is distilled into easy-to-understand executable actions which will allow you to win "more than your fair share." Ted's wisdom, garnered over many years of technical sales, biz-dev and marketing are shared in this book. I wish I had High-Tech Marketing Simplified early in my career!"

-Steve Donovan. VP of Sales at several semiconductor companies

"I've spent my career in technology communications. In his book, High-Tech Marketing Simplified, Ted condenses down a career of knowledge and experience into a very readable book. He helps companies get started in the market; while also being a resource for seasoned professionals looking to refine their approach. High-Tech Marketing Simplified is a great read and marketing tool for any high-tech professional seeking to improve their sales and marketing skills.
    - Allison DeLeo. Senior Vice President, Racepoint Global

"Ted has combined real world experience and helpful insights, all while explaining the key marketing points in a simplified approach. High-Tech Marketing Simplified shares a career of marketing and sales knowledge that everyone can implement. I highly recommend this book for high-tech marketing, business development and sales professionals."
    -Esam Elashmawi. Chief Marketing and Strategy Office, Lattice Semiconductor

"High-Tech Marketing Simplified clearly explains in a step-by-step manner how to execute an entire marketing cycle for semiconductors and high-tech products. Ted's shared experience is a gift for every organization marketing in the technology space."
    - Calista Redmond. CEO, RISC-V International

ABOUT THE AUTHOR

Ted J Marena is a recognized high-tech marketing and business development executive

www.ingramcontent.com/pod-product-compliance
Lightning Source LLC
Chambersburg PA
CBHW071422210326
41597CB00020B/3612